Kornelia Schriebl

Characterization of recombinant human Epo-Fc fusion protein

Kornelia Schriebl

Characterization of recombinant human Epo-Fc fusion protein

Focused on product quality aspects

Südwestdeutscher Verlag für Hochschulschriften

Impressum/Imprint (nur für Deutschland/ only for Germany)
Bibliografische Information der Deutschen Nationalbibliothek: Die Deutsche Nationalbibliothek verzeichnet diese Publikation in der Deutschen Nationalbibliografie; detaillierte bibliografische Daten sind im Internet über http://dnb.d-nb.de abrufbar.
Alle in diesem Buch genannten Marken und Produktnamen unterliegen warenzeichen-, marken- oder patentrechtlichem Schutz bzw. sind Warenzeichen oder eingetragene Warenzeichen der jeweiligen Inhaber. Die Wiedergabe von Marken, Produktnamen, Gebrauchsnamen, Handelsnamen, Warenbezeichnungen u.s.w. in diesem Werk berechtigt auch ohne besondere Kennzeichnung nicht zu der Annahme, dass solche Namen im Sinne der Warenzeichen- und Markenschutzgesetzgebung als frei zu betrachten wären und daher von jedermann benutzt werden dürften.

Verlag: Südwestdeutscher Verlag für Hochschulschriften Aktiengesellschaft & Co. KG
Dudweiler Landstr. 99, 66123 Saarbrücken, Deutschland
Telefon +49 681 37 20 271-1, Telefax +49 681 37 20 271-0, Email: info@svh-verlag.de
Zugl.: Vienna, University of Natural Resources and Applied Life Sciences, Diss., 2006

Herstellung in Deutschland:
Schaltungsdienst Lange o.H.G., Berlin
Books on Demand GmbH, Norderstedt
Reha GmbH, Saarbrücken
Amazon Distribution GmbH, Leipzig
ISBN: 978-3-8381-0400-3

Imprint (only for USA, GB)
Bibliographic information published by the Deutsche Nationalbibliothek: The Deutsche Nationalbibliothek lists this publication in the Deutsche Nationalbibliografie; detailed bibliographic data are available in the Internet at http://dnb.d-nb.de.
Any brand names and product names mentioned in this book are subject to trademark, brand or patent protection and are trademarks or registered trademarks of their respective holders. The use of brand names, product names, common names, trade names, product descriptions etc. even without a particular marking in this works is in no way to be construed to mean that such names may be regarded as unrestricted in respect of trademark and brand protection legislation and could thus be used by anyone.

Publisher:
Südwestdeutscher Verlag für Hochschulschriften Aktiengesellschaft & Co. KG
Dudweiler Landstr. 99, 66123 Saarbrücken, Germany
Phone +49 681 37 20 271-1, Fax +49 681 37 20 271-0, Email: info@svh-verlag.de

Copyright © 2009 by the author and Südwestdeutscher Verlag für Hochschulschriften Aktiengesellschaft & Co. KG and licensors
All rights reserved. Saarbrücken 2009

Printed in the U.S.A.
Printed in the U.K. by (see last page)
ISBN: 978-3-8381-0400-3

The content of this book has been slightly modified from the original work.

ACKNOWLEDGEMENTS

Thanks to Prof. Karola Vorauer-Uhl for supervising my work. Over the years, she was someone whom I could turn to for a word of advice. Interacting with her has taught me how not to lose the big picture and her pinpoint accuracy helped me to have an eye on the finer details.

The colleagues at the institute and particularly the Analytics/Liposomology group made my time at the IAM thoroughly enjoyable.

I acknowledge gratefully the funding from the Austrian Center of Biopharmaceutical Technology (ACBT).

Special thanks to my family, my boyfriend Oliver and in particular to my mum who went to great lengths to make all of this possible.

CONTENTS

1 SUMMARY 3

2 INTRODUCTION 5

   2.1 FROM GENE TO RECOMBINANT EXPRESSION CELL LINE 5

   2.2 PRODUCT QUALITY ANALYSIS 7

      2.2.1 Molecular weight determination 8

      2.2.2 Glycoprotein separation by charge 9

      2.2.3 Glycoprotein detection 9

      2.2.4 Oligosaccharide analysis 10

3 GOAL 13

4 DISCUSSION 15

5 ABBREVIATIONS 19

6 REFERENCES 20

7 VALIDATION REPORT 25

   7.1 INTRODUCTION 25

   7.2 MATERIAL AND METHODS 26

      7.2.1 Chemicals 26

      7.2.2 Equipment 26

      7.2.3 Standard solutions 26

      7.2.4 Enzymatic assay 27

   7.3 VALIDATION PARAMETER 28

      7.3.1 Specificity 28

      7.3.2 Linearity, range and variance homogeneity 28

- *7.3.3 Limit of detection and quantitation* 28
- *7.3.4 Accuracy* 29
- *7.3.5 Precision* 29
- *7.3.6 Robustness* 29
- 7.4 APPRAISAL FACTORS 29
  - *7.4.1 Acceptance criteria* 29
- 7.5 RESULTS 30
  - *7.5.1 Specificity* 30
  - *7.5.2 Linearity* 30
  - *7.5.3 Range* 32
- 7.6 VARIANCE HOMOGENEITY 33
  - *7.6.1 Limit of detection and quantitation* 33
  - *7.6.2 Accuracy* 33
  - *7.6.3 Precision* 36
  - *7.6.4 Robustness* 38
- 7.7 DISCUSSION 41
- 7.8 REFERENCES 42
- 7.9 FORMULARY 43

8 PUBLICATIONS 44

1 Summary

One major goal in modern biotechnological research and development is to reduce the development time until registration to increase specific productivity and to advance product quality. In the past, big efforts were made to increase the volumetric yield by optimization in fermentation and downstream technologies. However, polypeptides used as therapeutics get more complex, thus not only the production yield is important but also optimal product quality is required. As an example, the human recombinant erythropoietin, a highly glycosylated hormone, is shown as one product by which the quality is more important than the volumetric yield. Based on the specified product characteristics only 10-20% of the expressed protein achieves the high quality level, thus this circumstance leads to high product costs. As shown by this example, not only technology aspects must be taken into account, but also in an even more complex manner analytical expertise is necessary. Therefore, an analytical platform including sensitive, robust and reproducible methods is the basis for a successful product development strategy. In an ideal case, these methods are applicable in clone selection, optimization of fermentation and downstream processing, as well as for quality control affairs. The industrial processes, as mentioned above, can only be strengthen by extensive and systematic studies based on scientific research. To change the strategy from trial and error models to more systematic, universal models, is one of the challenges in biotechnology. For this reason, we have chosen our model-protein recombinant human Epo-Fc (rhEpo-Fc) to study cell physiology aspects, product maturation as well as the specific cell capacity itself.

The aim of the present PhD thesis was to establish respectively, optimize an analytical platform which is suitable to characterize our model protein, expressed in CHO cells. In addition, these methods were used to analyze the secreted product during cell screening, to assist the clone selection and for process development. Analytical methods for product quantification by ELISA, a microtiterplate assay for the quantification of N-acetylneuraminic acid (Neu5Ac), a quantitative assay for the isoform pattern, based on IEF and a method to quantify the amount of oligosaccharides were developed.

The determination of the volumetric concentration was performed by a classical ELISA method to analyze many samples with various concentrations. The quantification of the sialic acid amount was also performed in microtiterplates. This system is advantageous to analyze many samples in parallel with increased reproducibility. To make this method fit for quality control, validation according the ICH guidelines was performed. Isoelectric focusing for the determination of complex glycoproteins is state of the art.

Nevertheless, for quantification of the isoform pattern, as required for a registered product, the choice of the staining procedure is essential. For this approach a panel study with different fluorescence dyes were performed to evaluate parameters such as linear, dynamic ranges, reproducibility and last but not least, possible application in industrial use. Our optimized IEF, using CyDyes for prestaining, affords the analysis up to three proteins in one lane and additionally, proteins in protein free medium can be directly analyzed without further purification. The first advantage can be used to analyze samples and reference standards simultaneously and the second advantage is given by reduced analytical operating expense. Beside this basic characterization more intensive analytical work is necessary to obtain information of the glycan pattern, because no non-destructive method is known. Particularly, quantitative analysis, which is required for product specification and for batch to batch consistency efforts high scientific methods. Furthermore, in the special case of rhEpo-Fc, a separation must be established to analyze the rhFc part separately from the hormone part.

In conclusion, during this work, we have established a set of generally suitable techniques for protein characterization of complex glycoproteins. However, specific optimization for the protein of interest is unavoidable. Because, posttranslational modification of complex polypeptides are determined by host-cell-protein-configuration, culture environment as well as process design, only research networks can successfully develop advanced technology platforms, wherein analytical know how offers a valuable part.

2 Introduction

The major drawbacks of biopharmaceutical product development are long development time (around 10 years), high costs and high risks. On average, one out of ten products in the development pipeline reaches the market. Therefore, cell respectively clone screening must be established and monitored early for successful product development.

2.1 From gene to recombinant expression cell line

Production of recombinant proteins in animal cell culture for research and development as well as commercial purposes starts with the construction of an applicable cell line. The choice of expression systems is generally based on the nature of the protein to be expressed, the quantity of material needed and the period of time over which maintained production is required. Mammalian expression systems are generally used for the production of complex and/or glycosylated proteins. Because these cell lines are able to produce glycoproteins that are similar to human glycoproteins. For production of small quantities over a short period, transient expression systems can be the method of choice. These systems are mainly used for research, because cell lines usually loses its ability to express the target protein in time and the heterogeneous DNA is usually not integrated into the host genome [1]. However, for commercial protein production a stable expression system with the capability of continuous and stable product expression over an extended period is required. But a stable cell line, in which the heterogeneous DNA is integrated into the host genome and is maintained throughout many generations, requires an extensive time (approx. 6 months) to be developed and to demonstrate that it has the required properties of clonality, integration of the target gene into the genome and stability of high and continuous expression.

Various cell lines are capable for high level expression of recombinant proteins. The most commonly used cell lines are Chinese hamster ovary (CHO) cells, baby hamster kidney (BHK) cells, human embryonic kidney cells (HEK 293) and different myeloma cell lines such as NS0, SP2 and YB2/0. Depending on the protein of interest respectively product characteristics a suitable host cell line is chosen [2-4]. Cell lines such as CHO, BHK and NS0 indicated that the complex carbohydrate structures, which are produced by the corresponding cell line, contain similar oligosaccharide structures naturally occurring on human proteins. Murine cell lines, unlike hamster lines, may produce significant quantities of terminal *N*-glycolylneuraminic acid in place of terminal *N*-acetylneuraminic acid, or display an α-1,3-galactosyltransferase activity, that may lead oligosaccharide structures which are potentially immunogenic in humans. In contrast, CHO and BHK cells lack the ability to produce certain structures normally found on human glycoproteins, including terminal sialic acid residues attached in an α-2,6 linkage to galactose or a bisecting GlcNAc residue as well as an α-1,3/4 attached fucose [5].

Apart from the host cell line, the ability to achieve high-level mRNA expression appears to depend primarily on selecting and optimizing the right combination of host cells, expression vector and selection strategy. For instance, various dihydrofolate reductase (DHFR) deficient CHO cells are available. These cells can be combined with a spectrum of vectors that combine promoters highly active in CHO cells with strategies to accomplish efficient amplification using DHFR gene replacement [6,7]. During the initial selection process and/or following additional rounds of amplification, individual clones with high specific productivity are isolated and characterized. Detailed product characterization should be implemented after final amplification of potential production clones, because during amplification many clones are not stable, thus product characterization is not efficient. In order to evaluate the specific productivity, which is the amount of product produced per cell per unit time (pmol/cell/day), of hundreds of clones, a high throughput assay is needed. For instance, the accumulated product in cell culture supernatants can be evaluated using an immunoassay (ELISA). A robust ELISA is suitable in view of its high sensitivity and high throughput aspects. The cell density can be obtained by optical measurement, either of light transmission through the cells themselves or of a dye that identifies viable cells.

In the case of an expressed protein, where no standard compound is available, clones can be screened by SDS-PAGE using an fluorescent dye for semi-quantitative determination. Furthermore, in some cases clones can be screened by quantitative mRNA analysis using real-time PCR [1]. But in such cases where the protein secretion is limited for instance by translation, real-time PCR screening may not be successful.

On the basis of these results, clones with good growth rates and high specific production properties will be selected. Selected clones are generally expanded on tissue culture flasks until enough cells are available to inoculate them into a suspension culture. Depending on the cell type, two T-175 tissue culture flasks usually contain sufficient cells to inoculate a spinner flask with a 200 ml working volume with ~ 3–4x10^5 cells/ml [1].

2.2 Product quality analysis

After cell adaptation to suspension and/or serum-free conditions, their growth properties are much closer to that in large-scale cell culture process than in small scale tissue culture during clone development. At this stage detailed product quality analyses such as analysis of molecular weight, of sialic acid amount, of isoform pattern and analysis of oligosaccharide structures can be monitored. Figure 1 shows the strategy from transfection until production, schematically.

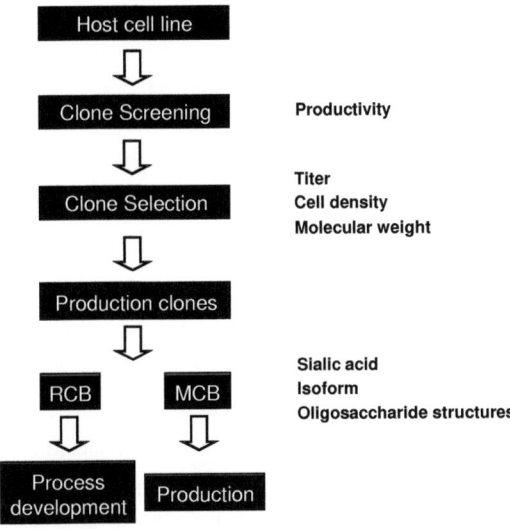

Figure 1: Flow chart of analytical aspects during clone screening.

Product quality analyses during these steps are very important to ensure that the final selected clone has the most desirable cell culture and product properties. Therefore, titer, cell density and molecular weight determinations are performed during clone selection. But it should be ensured that detailed product quality determination such as sialic acid content, isoform pattern and oligosaccharide structure analyses are performed from final amplified clones. Because the unstable the clone, the unconfident the result.

A breadth of work during the last decade has demonstrated how various reactions in the glycosylation pathway can be influenced by cell culture environment, host cell line selection and protein specific features [4,8,9] leading to production of molecules with variable or suboptimal clearance or bioactivity *in-vivo*.

As shown in Figure 2, various methods are available for protein characterization.

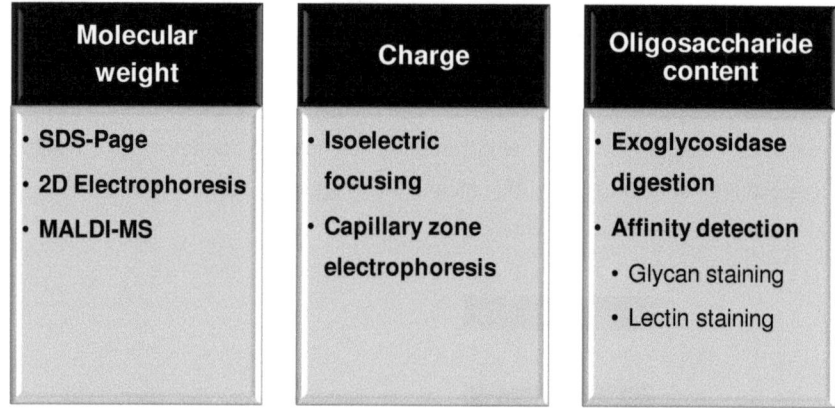

Figure 2: Analysis techniques to characterize proteins, respectively glycoproteins.

2.2.1 Molecular weight determination

The molecular weight of a protein and/or glycoprotein can be analyzed by SDS-PAGE, 2D-Electrophoresis and matrix-assisted laser desorption ionization mass spectrometry (MALDI-MS). In the case of SDS-PAGE, the proteins are separated by molecular weight. By loading with the anionic detergent sodium dodecylsulphate (SDS), the charge of the proteins is almost masked resulting in anionic micelles with a constant net charge per unit. In addition, the tertiary and secondary structures are cancelled due to the disruption of the hydrogen bonds and unfolding of the molecules. SDS-PAGE is also a method accepted by the FDA to determine the purity and quantity of recombinant proteins, to compare material from different batches and to monitor product quality during a batch [10,11].

2D-Electrophoresis can be used to detect posttranslational modifications and/or molecular weight changes during clone screening and fermentation [12]. The protein is separated by p*I* and molecular weight. Thus, each glycoprotein is visualized as a series of isoforms which occurs of variable size and charge densities.

Molecular weight can also be measured by MALDI-MS. It was developed to analyze large molecules (< 300 kDa) [13]. Generally, the sample is embedded in a low-molecular weight UV-absorbing matrix. During analysis the sample is ionized by a pulsed laser. Very few fragments result, thus MALDI-MS can be used for screening molecular ions for high throughput and high sensitivity [14,15].

2.2.2 Glycoprotein separation by charge

The presence of different oligosaccharides can be used to separate the glycoforms by charges, which are presented by the negatively charged sialic acid residues.

On isoelectric focusing (IEF) gels, glycoproteins were separated according their p*I* values. A pH gradient in the gel is created by ampholyte molecules. The glycoprotein migrates to the zone of the gel where the p*I* of the glycoprotein matches the pH in the gel [12,16]. This method is semi-quantitative regarding to the staining procedure and can be used to detect macroheterogeneity changes of glycoproteins during clone screening or fermentation [3]. The analysis of microheterogeneity is more demanding, because glycans themselves do not result in different isoforms.

A newer technique to detect glycoform heterogeneity is capillary zone electrophoresis (CZE). This technique separates molecules according their different charge to mass ratio. However, migration times of the different isoforms are very similar. Thereby, minimal shifts during the analysis may negatively interfere with isoform classification. This might hamper the correct assignment of each of the isoforms. Different approaches to improve accuracy in band assignment in CZE have been published. Some of these approaches use internal standards spiked into samples to improve the identification. The migration parameter value of each isoform of a sample, analyzed under given conditions, was compared with the reference values to identify the isoform [17]. Generally, due to the complex interactions of the glycoprotein with the capillary wall, each separation needs to be optimized to improve resolution by changing the pH, the buffer type and the organic modifiers [18].

2.2.3 Glycoprotein detection

Another method to verify glycosylation is to use a specific glycan stain or specific lectins after SDS-PAGE and/or Western blotting.

In the case of glycan staining, the glycoproteins are separated by SDS-PAGE and visualized with a sensitive green-fluorescent glycoprotein specific stain, named Pro-Q Emerald 488 dye that allows detection of glycoproteins on PVDF membranes or directly in polyacrylamide gels [19]. Pro-Q Emerald 488 dye may be conjugated to glycoproteins by a periodic acid Schiff´s (PAS) mechanism using very mild reaction conditions at room temperature. Competitive post-staining may be done with other protein specific fluorescent dyes such as Sypro Ruby and Deep Purple. Thus, detection of glycosylated and non-glycosylated proteins on the same gel is feasible.

Lectins represent another approach for glycoprotein detection. Generally, a lectin is a protein which binds certain carbohydrates but is neither an enzyme nor an antibody. Thus, the practical relevance of lectin based analysis depends on the location of the carbohydrate moieties and therefore the steric accessibility of the protein, respectively binding capacity of certain lectins [20].

Furthermore, glycoproteins can be characterized by exoglycosidases. These enzymes are able to cleave specific oligosaccharide bonds on a glycoprotein, thus reducing the molecular weight of the protein. Exoglycosidases can be used to determine the identity, absolute and anomeric configuration and with some enzymes the linkage position of the oligosaccharide [16,21]. Thus, the degree and type of glycosylation can be determined by comparing the cleaved protein to the original glycoprotein [15,22].

2.2.4 Oligosaccharide analysis

Oligosaccharides can be isolated from glycoproteins by chemical or enzymatic cleavage [15,22,23]. Hydrozinolysis is the most common chemical method for cleaving *N*- and *O*-linked oligosaccharides non-selectively. Most oligosaccharides can be cleaved, but some reducing-end sugar moieties may be modified [3,24]. The *N*-linked oligosaccharides can be released by the enzyme *N*-glycosidase F (PNGase F) [25,26]. For the release of *O*-linked oligosaccharides no suitable enzymatic method is available. Other endoglycosidases can be used to cleave specific bonds in the oligosaccharide or between the oligosaccharide and protein [15,22,23]. Once the oligosaccharide is removed from the glycoprotein, an oligosaccharide profile can be analyzed. Often fluorophores are used for oligosaccharide labeling to improve sensitivity [27]. Oligosaccharides can be labeled with the tag of choice, such as 2-aminobenzoic acid, 2-aminobenzamide, 2-aminopyridine and 2-aminoacridone. All tags use anhydrous DMSO-acetic acid (15–30 % (v/v) or acetic acid alone as the solvent for reductive animation and reaction temperatures of 65–90°C for 1-2 h or 35°C overnight in the presence of sodium cyanoborohydride, to give stable highly fluorescent derivatives [28-30].

The isolated, oligosaccharides are analyzed by methods such as FACE, HPLC, FAB-MS, ES-MS, MALDI-MS, LC-MS, NMR. Whereas labeled oligosaccharides are primarily analyzed by FACE and HPLC. Using these methods, mass, composition, linkages and sequence information of the oligosaccharides are obtained.

Fluorophore-assisted carbohydrate electrophoresis (FACE) is a relative simple and inexpensive technique. Oligosaccharides are labeled with 8-aminonaphthalene-1,3,6-trisulfonate (ANTS) and separated by mass on polyacrylamide gels for quantification

[15,31,32]. FACE allows direct comparison of samples for quantification, since multiple samples can be run on a single gel. This method can be used to monitor glycosylation changes during fermentation due to bioprocess changes [12].

HPLC methods are commonly used to characterize the glycosylation of the glycoprotein. For instance, normal-phase HPLC with an optimized acetonitrile–ammonium acetate gradient on a TSK-Gel Amide-80 column is used [33]. The polar functional groups of the Amide-80 column interact with the hydroxyl groups on the oligosaccharide, thus labeled neutral and charged oligosaccharides can be separated simultaneously on the basis of hydrophilic interaction. In principle, large oligosaccharides are eluted by increasing concentration of ammonium acetate. For this reason, neutral oligosaccharides are eluted first, followed by charged oligosaccharides. The N-glycans are identified by the retention times of defined oligosaccharide standards and by sequential exoglycosidase cleavage. Thus, macro- and microheterogeneity can be determined.

Mass spectrometry can be used to determine the primary structure of a glycoprotein, like branching, linkages, configuration and identification of isomer sugars. Liquid chromatography coupled to mass spectrometry (LC-MS) as well as tandem MS-MS methods can also improve oligosaccharide analysis [23,34,35].

In conclusion, many analytical techniques are available to determine and quantify protein glycosylation during clone screening and further production steps. These analytical techniques are highly sensitive and able to detect and separate large molecules with very small differences. However, to entirely characterize a glycoprotein, at least two or more orthogonal analytical techniques are needed.

Recent studies have demonstrated that specific alteration of oligosaccharide structures on a recombinant protein can be achieved by overexpression of appropriate glycosyltransferases. This strategy can be used either to enhance glycan quality by increasing the homogeneity of native structures or introducing non host cell residues to specialize glycan quality and function. For example, the overexpression of galactosyltransferase and sialyltransferase in CHO cells has been shown to lead to corresponding increases in the galactose and sialic acid levels on recombinant therapeutic protein expression by these cells [36]. Other groups have successfully overexpressed N-acetylglucosaminyltransferase III in order to increase the fraction of bisecting N-acetylglucosamine residues on antibodies produced in CHO cells [37,38] or introduced sialic acid in an α-2,6 linkage to glycoproteins synthesized by CHO and BHK cells that lack the specific sialyltransferase [39-41].

However, what is a good product quality? Dependent on the point of view we have to differentiate between quality required by the regulatory (European Pharmacopoeia) and scientific product quality. From the regulatory point of view, product quality is exactly defined. For instance, in the case of the hormone erythropoietin the amount of sialic acid/molecule erythropoietin, the isoform pattern, the verification of dimers and related substances of higher molecular mass and the bioactivity of the product itself has to be determined with defined assays. In contrast to the required regulatory specifications, the scientific product quality does not have restrictions. It is used to analyze quality aspects of proteins, for instance fusion proteins, which are not found in nature.

In conclusion, protein glycosylation is the most common posttranslational modification in eukaryotic cells. 0.5–1% of the translated mammalian genome participates in oligosaccharide production and function [42]. Therefore, extensive determination of product quality during early stages of clone development is required.

3 Goal

Various recombinant proteins expressed in mammalian cells are used as biopharmaceuticals in immune suppression and replacement therapies, as well as diagnostic purposes. Therefore, efficient high quality products need to be produced. In the past, the major concern in product development was to reach high volumetric titers and to establish efficient fermentation and downstream processing technologies. Nowadays, the proteins produced for therapeutic application are more complex, in terms of posttranslational modifications, thus not only product quantities are important, even more quality aspects are relevant. One example of a complex, highly glycosylated protein is the human erythropoietin, which consists of 40% glycans in a specific configuration. Economic production of this product is only possible when optimized host-cell-protein systems, optimized process technologies and suitable analytical procedures are available.

Therefore, analytical development is required to establish and evaluate techniques to improve more methods for quality monitoring in cell screening, which is also suitable in process control and quality control.

For this reason, the goal of this work was to establish an analytical platform to recognize differences in product quality during clone development of the model protein recombinant human Epo-Fc (rhEpo-Fc). Beside the quantification of the product with an appropriate ELISA, the sialic acid amount, isoform pattern as well as the related oligosaccharide structures should be analyzed.

In principle, all these methods are commonly used for certain product characterization. But, in contrast to the known technologies, our goal was to establish methods enough sensitive, robust, but also reproducible to be used as high throughput assays, suitable for very small sample amounts with low product concentrations and whenever possible, without prior product purification. In addition, all methods were evaluated in terms of quantitative data. However, when purification is necessary, our model protein could be captured by Protein A affinity, either by small scale chromatography or alternatively, with Protein A beads.

For the quantification of the sialic acid amount, an enzymatic microtiterplate assay for the determination of N-acetylneuraminic acid should be established and validated according to the ICH guidelines for validation of analytical procedures.

For the determination of isoform pattern, directly performed with the culture supernatant, a quantitative method for the determination of specific isoforms should be developed.

After optimization of the p*I* range of the polyacrylamid gel, different fluorescent staining methods such as Deep Purple, Sypro Ruby and CyDye flours are evaluated. Reproducibility should be evaluated, using the most efficient staining method.

For further product analysis, *N*-glycan profiling using normal-phase HPLC of fluorescent labeled glycans should complete this analytical platform, where the two parts of the fusion protein should be analyzed separately.

4 Discussion

Recent studies have demonstrated that clone, respectively product screening must be established and monitored very early during product development to select efficient candidates for production. In this case, the product quantification alone is not the method of choice, thus also individual quality criteria must be tested. But in early stages of clone development two major problems are obvious. The first one is to define the optimal testing set-point and the second is, that during the initial selection process as well as the final round of amplification just very small sample amounts with low product concentrations are available.

For instance, during cell screening in the 96 well systems only 200 µl culture supernatant with a concentration of some pmol/ml are available for product quality analysis. For that purpose, neither purification nor detailed oligosaccharide analyses are possible. Assays which can be performed directly from culture supernatant are desirable, but the presence of interfering host cell compounds, such as serum proteins, prevents reliable results. Furthermore, immediately after transfection many clones are not stable, thus expensive analysis is not efficient.

However, for the analysis of stabilized clones we established a technique for quantitative isoform pattern analysis directly performed in serum-free culture supernatant.

In our approach, we evaluated the use of different CyDyes regarding reproducibility, suitability for crude cell-free culture supernatants and overlaying method. Our results, obtained with four independent analyses, showed a maximum SD of ± 0.9% corresponding isoforms. Based on these results, we analyzed different rhEpo-Fc samples. In this regard, purified as well as non-purified samples were analyzed. For evaluation, different culture supernatants containing rhEpo-Fc were harvested and a part of it was purified by affinity chromatography. All different culture supernatants were generated in serum-free media but under different culture conditions. The isoform pattern of the serum-free culture supernatant and the corresponding purified protein of all three analyzed samples showed comparable results. The maximum deviation between the corresponding isoforms of all three samples ranged between ± 0.4 – 1.2%. Based on these results, we could demonstrate that this technique, including 2-D Clean Up Kit and labeling with CyDye fluors, is suitable for protein characterization directly from protein free culture supernatants. In the case of rhEpo-Fc, the detection limit was 4 pmol. In fact, 100 µl of an rhEpo-Fc sample with a concentration of 500 pmol/ml was used for 2-D Clean Up Kit purification, whereas the protein pellet was dissolved in 8 µl resuspension solution.

Depending on the characteristics of the protein of interest as well as the number of isoforms, the detection limit is variable. For example, the detection limit of a protein with 4 isoforms will be lower in contrast to a protein with 20 isoforms (rhEpo-Fc). By the use of non-purified protein solutions, in combination with fluorescence techniques, faster analysis of protein isoforms in cell screening and fermentation is possible, avoiding prior time-consuming protein purification.

Furthermore, we directly compared the serum-free culture supernatant and the internal control sample by labeling them with different CyDye fluors which were separated on the same lane and scanned at the appropriate wavelengths [43]. With this technique up to three samples can be focused on the same lane under identical electrophoretic conditions. A fundamental benefit of this overlaying technique is the ability to co-detect and compare each sample in-gel with an internal standard. This internal standard can be used for normalization of the isoforms across all gels. With this approach, the experimental variation is further reduced and the accuracy of quantification is increased.

In many glycosylated proteins, the more or less structured glycans are terminated with a N-acetylneuraminic acid residue, to protect the protein from the capture by receptors in the liver, leading to decreased biological half life *in-vivo* [44,45]. Thus, methods are needed to quantify them as quality aspects. Most of the commonly used methods are not suitable for clone screening approach, because high concentrations and large volumes are necessary [46].

To circumvent the problem, we have developed a microtiterplate assay for the determination of sialic acid based on the already known principles of enzymatic methodology for the characterization of purified samples. The basis of this enzymatic assay is the hydrolysis of the glycoprotein-bound sialic acid by neuraminidase to release it as free compound. Thereafter, N-acetylneuraminic acid aldolase is used to convert the free sialic acid quantitatively to pyruvic acid and N-acetylmannosamine. Pyruvic acid can be assayed using enzymes such as lactic dehydrogenase, coupled to NADH oxidation, to reduce it. NADH oxidation to NAD^+ can be quantified spectrofluorometrically (Ex 340 nm, Em 465 nm). The parameters specificity, linearity, accuracy, precision, robustness, limit of detection and quantitation were studied according to the International Conference on Harmonization guidelines for analytical procedures [47]. The enzymatic assay is quantifiable in the range of 2 – 10 nmol/ml (64 – 320 pmol/well) sialic acid with a first order equation. In contrast, for the sialic acid assay of the European Pharmacopoeia a 10 fold higher product concentration is needed [46]. The validation documents show that this micro method is robust, accurate and reproducible for the intended purpose.

The proposed method is suitable for the analysis of purified samples for instance for routine quality analysis as well as clone screening.

For further characterization of each glycoprotein, oligosaccharide structure analysis must be performed. By this analysis, the product quality of selected clones can be determined as well as evaluation of process development can be monitored. All commonly used methods require more product amount than methods described above. As lower the oligosaccharide rate of each protein as much protein must be available. Nowadays no method is known to analyze the glycan structures directly on the protein. Prior to analysis, oligostructures must be removed from the protein-core [15,22,23], followed by separation [33] and in most cases labeling must be performed [27]. Thus, these methods are quite time consuming and expensive. Nevertheless, each product must be analyzed according the glycans and for biopharmaceuticals not only a qualitative analysis must be done, even more quantitative glycan pattern must be documented. Therefore, the intention in our project was not to establish a new glycan structure methodology, but to establish a quantifiable method and to analyze both fusion partners separately.

The *N*-linked oligosaccharides of the rhEpo-Fc fusion protein were released enzymatically [25,26] and analyzed by normal-phase HPLC [33]. *N*-glycans were analyzed from the rhEpo-Fc molecule as well as from the rhEpo and rhFc subunit, obtained by papain digestion [48]. The main detected *N*-glycans of the rhEpo molecule were tetrasialo tetraantennary fucosylated (34%), trisialo triantennary (20%) and monosialo biantennary fucosylated (14%). These findings are also consistent with the results of other studies, which have shown that the majority of the *N*-glycans of rhEpo were tetraantennary, but they also show that not all of them were fully sialylated [49,50]. The corresponding rhFc molecule consisted predominantly of monogalacto biantennary fucosylated (24%), monogalacto biantennary (20%) followed by asialo agalacto biantennary fucosylated, asialo biantennary and asialo biantennary fucosylated (each 12%) structures. These findings are consistent with the results of other studies for the characterization of antibody glycosylation, which have shown that the majority of the *N*-glycans were asialo biantennary fucosylated, monogalacto biantennary fucosylated and agalacto biantennary fucosylated [51,52].

In conclusion, the goal of this work was to establish an analytical platform to recognize differences in protein quality during clone selection and process development. In the 96 well system product quality, molecular weight determination by SDS-PAGE and Western blotting could be obtained.

By our studies we could demonstrate that some methods are suitable for samples with low volume paired with low concentration, while others are not such sufficient.

In the case of serum containing samples, most of the analytical methods cannot be directly performed in the supernatant, because much interference is predictable, thus purification is a must. Therefore, analysis can only be performed if enough material is available for purification and analyses such as sialic acid amount, isoform pattern and oligosaccharide structure. This circumstance complicates all activities necessary in optimal clone screening. Nevertheless, if a fast decision during clone screening is needed, product concentration determination (ELISA) as well as isoelectric focusing including 2-D Clean Up Kit are the methods of choice. If a longer decision period is possible (about 1 - 2 weeks), purification and sialic acid determination should be performed. Oligosaccharide analysis will give the most detailed information about the glycans, but the analysis is very time-consuming (2 weeks). This technique should only be used for pre-selected clones. Pre-selection can be performed by isoelectric focusing and sialic acid determination. Different to serum transfection technique, protein free transfection protocols are advantageous concerning faster reliable analysis. In this case, for the isoform determination no purification is necessary, thus by methods using sensitive labeling, fast analysis can be performed. Our established method, using different CyDye fluors is an excellent example, because if an appropriate standard of the certain product is available, clone selection can be performed with this overlaying technique. Furthermore, if unavoidable, a less complex purification procedure for protein free supernatants is possible, which reduces time and costs. In combination with more sensitive, but also valid methods, decisions can be accelerated.

5 Abbreviations

ANTS	8-aminonaphthalene-1,3,6-trisulfonate
BHK	Baby hamster kidney
CZE	Capillary zone electrophoresis
CHO	Chinese Hamster ovary
DHFR	Dihydrofolate reductase
ELISA	Enzyme linked immunosorbent assay
ES-MS	Electrospray Mass Spectrometry
FAB-MS	Fast-atom bombardment mass spectrometry
FACE	Fluorophore-assisted carbohydrate electrophoresis
FDA	Food and Drug Administration
GlcNAc	*N*-acetylglucosamine
HEK	Human embryonic kidney
HPLC	High performance liquid chromatography
ICH	International Conference on Harmonization
IEF	Isoelectric focusing
LC-MS	Liquid chromatography mass spectrometry
MALDI-MS	Matrix-assisted laser desorption ionization mass spectrometry
MCB	Master cell bank
NADH	β-Nicotinamide-Adenine Dinucleotide
NMR	Nuclear Magnetic Resonance
p*I*	Isoelectric point
RCB	Research cell bank
rhEpo	recombinant human Erythropoietin
RT-PCR	Real-Time Polymerase Chain Reaction
SD	Standard deviation

6 References

[1] S.S. Ozturk and W.S. Hu, *Cell Culture Technology for Pharmaceutical and Cell-based Therapies*, CRC Press, New York, 2006.

[2] P.K. Bhatia and A. Mukhopadhyay, *Protein glycosylation: implications for in vivo functions and therapeutic applications*. Adv. Biochem. Eng. Biotechnol. 64 (1999) 155-201.

[3] N. Jenkins and E.M. Curling, *Glycosylation of recombinant proteins: problems and prospects*. Enzyme. Microb. Technol. 16 (1994) 354-364.

[4] E. Grabenhorst, P. Schlenke, et al., *Genetic engineering of recombinant glycoproteins and the glycosylation pathway in mammalian host cells*. Glycoconj. J. 16 (1999) 81-97.

[5] J.P. Kamerling, *Carbohydrate features of recombinant human glycoproteins*. Biotecnol. Apl. 13 (1996) 167-180.

[6] A.D. Levinson, *Expression of heterologous genes in mammalian cells*. Methods Enzymol. 185 (1990) 485-487.

[7] C.R. Wood, A.J. Dorner, et al., *High level synthesis of immunoglobulins in Chinese hamster ovary cells*. J. Immunol. 145 (1990) 3011-3016.

[8] C.F. Goochee, M.J. Gramer, et al., *The oligosaccharides of glycoproteins: bioprocess factors affecting oligosaccharide structure and their effect on glycoprotein properties*. Biotechnology (NY) 9 (1991) 1347-1355.

[9] N. Jenkins, R.B. Parekh, et al., *Getting the glycosylation right: implications for the biotechnology industry*. Nat Biotechnol 14 (1996) 975-981.

[10] D.K. Robinson, C.P. Chan, et al., *Characterization of a recombinant antibody produced in the course of a high yield fed-batch process*. Biotechnology and Bioengineering 44 (1994) 727-735.

[11] T.P. Patel, R.B. Parekh, et al., *Different culture methods lead to differences in glycosylation of a murine IgG monoclonal antibody*. Biochem. J. 285 (1992) 839-845.

[12] M. Taverna, N.T. Tran, et al., *Electrophoretic methods for process monitoring and the quality assessment of recombinant glycoproteins*. Electrophoresis 19 (1998) 2572-2594.

[13] D.J. Harvey, *Matrix-assisted laser desorption/ionization mass spectrometry of carbohydrates*. Mass. Spectrom. Rev. 18 (1999) 349-450.

[14] A. Dell and H.R. Morris, *Glycoprotein structure determination by mass spectrometry.* Science 291 (2001) 2351-2356.

[15] H. Geyer and R. Geyer, *Strategies for glycoconjugate analysis.* Acta. Anat. (Basel). 161 (1998) 18-35.

[16] N.H. Packer and M.J. Harrison, *Glycobiology and proteomics: is mass spectrometry the Holy Grail?* Electrophoresis 19 (1998) 1872-1882.

[17] I. Lacunza, P. Lara-Quintanar, et al., *Selection of migration parameters for a highly reliable assignment of bands of isoforms of erythropoietin separated by capillary electrophoresis.* Electrophoresis 25 (2004) 1569-1579.

[18] K. Kakehi and S. Honda, *Analysis of glycoproteins, glycopeptides and glycoprotein-derived oligosaccharides by high-performance capillary electrophoresis.* J. Chromatogr. A. 720 (1996) 377-393.

[19] T.H. Steinberg, K. Pretty On Top, et al., *Rapid and simple single nanogram detection of glycoproteins in polyacrylamide gels and on electroblots.* Proteomics 1 (2001) 841-855.

[20] M. Liljeblad, A. Lundblad, et al., *Analysis of glycoproteins in cell culture supernatants using a lectin immunosensor technique.* Biosens. Bioelectron. 17 (2002) 883-891.

[21] S. Prime, J. Dearnley, et al., *Oligosaccharide sequencing based on exo- and endoglycosidase digestion and liquid chromatographic analysis of the products.* J Chromatogr A. 720 (1996) 263-274.

[22] R.A. O'Neill, *Enzymatic release of oligosaccharides from glycoproteins for chromatographic and electrophoretic analysis.* J. Chromatogr. A. 720 (1996) 201-215.

[23] T. Merry, *Current techniques in protein glycosylation analysis. A guide to their application.* Acta. Biochim. Pol. 46 (1999) 303-314.

[24] R.B. Parekh and T.P. Patel, *Comparing the glycosylation patterns of recombinant glycoproteins.* Trends Biotechnol. 10 (1992) 276-280.

[25] A.L. Tarentino and T.H. Plummer, Jr., *Enzymatic deglycosylation of asparagine-linked glycans: purification, properties, and specificity of oligosaccharide-cleaving enzymes from Flavobacterium meningosepticum.* Methods Enzymol. 230 (1994) 44-57.

[26] R.B. Trimble and A.L. Tarentino, *Identification of distinct endoglycosidase (endo) activities in Flavobacterium meningosepticum: endo F1, endo F2, and endo F3. Endo F1 and endo H hydrolyze only high mannose and hybrid glycans.* J. Biol. Chem. 266 (1991) 1646-1651.

[27] F.N. Lamari, R. Kuhn, et al., *Derivatization of carbohydrates for chromatographic, electrophoretic and mass spectrometric structure analysis.* J. Chromatogr. B. 793 (2003) 15-36.

[28] J.C. Bigge, T.P. Patel, et al., *Nonselective and efficient fluorescent labeling of glycans using 2-amino benzamide and anthranilic acid.* Anal. Biochem. 230 (1995) 229-238.

[29] F. Matsuura, M. Ohta, et al., *The combination of normal phase with reversed phase high performance liquid chromatography for the analysis of asparagine-linked neutral oligosaccharides labelled with p-aminobenzoic ethyl ester.* Biomed. Chromatogr. 6 (1992) 77-83.

[30] F. Chen, T.S. Dobashi, et al., *Quantitative analysis of sugar constituents of glycoproteins by capillary electrophoresis.* Glycobiology 8 (1998) 1045-1052.

[31] P. Jackson, *The use of polyacrylamide-gel electrophoresis for the high-resolution separation of reducing saccharides labelled with the fluorophore 8-aminonaphthalene-1,3,6-trisulphonic acid. Detection of picomolar quantities by an imaging system based on a cooled charge-coupled device.* Biochem. J. 270 (1990) 705-713.

[32] Friedman Y. and Higgins E. A., *A Method for Monitoring the Glycosylation of Recombinant Glycoproteins from Conditioned Medium, Using Fluorophore-Assisted Carbohydrate Electrophoresis.* Anal. Biochem. 228 (1995) 221-225.

[33] C.T. Yuen, P.L. Storring, et al., *Relationships between the N-glycan structures and biological activities of recombinant human erythropoietins produced using different culture conditions and purification procedures.* Br. J. Haematol. 121 (2003) 511-526.

[34] M. Ohta, N. Kawasaki, et al., *Usefulness of glycopeptide mapping by liquid chromatography/mass spectrometry in comparability assessment of glycoprotein products.* Biologicals 30 (2002) 235-244.

[35] N. Kawasaki, M. Ohta, et al., *Usefulness of sugar mapping by liquid chromatography/mass spectrometry in comparability assessments of glycoprotein products.* Biologicals 30 (2002) 113-123.

[36] S. Weikert, D. Papac, et al., *Engineering Chinese hamster ovary cells to maximize sialic acid content of recombinant glycoproteins.* Nat. Biotechnol. 17 (1999) 1116-1121.

[37] J. Davies, L. Jiang, et al., *Expression of GnTIII in a recombinant anti-CD20 CHO production cell line: Expression of antibodies with altered glycoforms leads to an increase in ADCC through higher affinity for FC gamma RIII.* Biotechnol. Bioeng. 74 (2001) 288-294.

[38] P. Umana, J. Jean-Mairet, et al., *Engineered glycoforms of an antineuroblastoma IgG1 with optimized antibody-dependent cellular cytotoxic activity.* Nat. Biotechnol. 17 (1999) 176-180.

[39] E. Grabenhorst, A. Hoffmann, et al., *Construction of stable BHK-21 cells coexpressing human secretory glycoproteins and human Gal(beta 1-4)GlcNAc-R alpha 2,6-sialyltransferase alpha 2,6-linked NeuAc is preferentially attached to the Gal(beta 1-4)GlcNAc(beta 1-2)Man(alpha 1-3)-branch of diantennary oligosaccharides from secreted recombinant beta-trace protein.* Eur. J. Biochem. 232 (1995) 718-725.

[40] L. Monaco, A. Marc, et al., *Genetic engineering of a2,6-sialyltransferase in recombinant CHO cells and its effects on the sialylation of recombinant interferon-g.* Cytotechnology 22 (1996) 197-203.

[41] E.U. Lee, J. Roth, et al., *Alteration of terminal glycosylation sequences on N-linked oligosaccharides of Chinese hamster ovary cells by expression of beta-galactoside alpha 2,6-sialyltransferase.* J. Biol. Chem. 264 (1989) 13848-13855.

[42] A. Varki and J.D. Marth, *Oligosaccharides in vertebrate development.* Sem. Dev. Biol. 6 (1995) 127-138.

[43] R. Tonge, J. Shaw, et al., *Validation and development of fluorescence two-dimensional differential gel electrophoresis proteomics technology.* Proteomics 1 (2001) 377-396.

[44] J.L. Spivak and B.B. Hogans, *The in vivo metabolism of recombinant human erythropoietin in the rat.* Blood 73 (1989) 90-99.

[45] M.N. Fukuda, H. Sasaki, et al., *Survival of recombinant erythropoietin in the circulation: the role of carbohydrates.* Blood 73 (1989) 84-89.

[46] European Pharmacopoeia 4, 1316 (2002) 1123-1128.

[47] International Conference on Harmonization, Draft Guideline on Validation of Analytical Procedures: Methodology, ICH-Q2A, 2B (1996)

[48] M. Adamczyk, J.C. Gebler, et al., *Papain digestion of different mouse IgG subclasses as studied by electrospray mass spectrometry.* J. Immunol. Methods 237 (2000) 95-104.

[49] K. Kanazawa, K. Ashida, et al., *Establishment of a method for mapping of N-linked oligosaccharides and its use to analyze industrially produced recombinant erythropoietin.* Biol. Pharm. Bull. 22 (1999) 339-346.

[50] C.H. Hokke, A.A. Bergwerff, et al., *Structural analysis of the sialylated N- and O-linked carbohydrate chains of recombinant human erythropoietin expressed in Chinese hamster ovary cells. Sialylation patterns and branch location of dimeric N-acetyllactosamine units.* Eur. J. Biochem. 228 (1995) 981-1008.

[51] R. Jefferis, J. Lund, et al., *A comparative study of the N-linked oligosaccharide structures of human IgG subclass proteins.* Biochem. J. 268 (1990) 529-537.

[52] G.R. Guile, P.M. Rudd, et al., *A Rapid High-Resolution High-Performance Liquid Chromatographic Method for Separating Glycan Mixtures and Analyzing Oligosaccharide Profiles.* Anal. Biochem. 240 (1996) 210-226.

7 Validation report

Validation of an enzymatic microtiterplate assay for the determination of N-acetylneuraminic acid

7.1 Introduction

The validation of the enzymatic microtiterplate assay for the determination of N-acetylneuraminic acid (Neu5Ac) was performed according to the ICH guidelines for validation of analytical procedures [1,2]. The microtiterplate assay was verified regarding to specificity, linearity, range, variance homogeneity, accuracy, precision and robustness.

The basis of this enzymatic assay is the hydrolysis of the glycoprotein- or glycolipid-bound sialic acid by neuraminidase to release the free compound. Thereafter, N-acetylneuraminic acid aldolase (NANA-aldolase) is used to convert the free sialic acid quantitatively to pyruvic acid and N-acetylmannosamine. During the final reaction, the pyruvic acid, generated in the second reaction, can be assayed using enzymes such as lactic dehydrogenase, coupled to NADH oxidation, to reduce pyruvic acid. NADH oxidation can be spectrophotometrically (Ex 340 nm, Em 465 nm) quantified (Fig. 1).

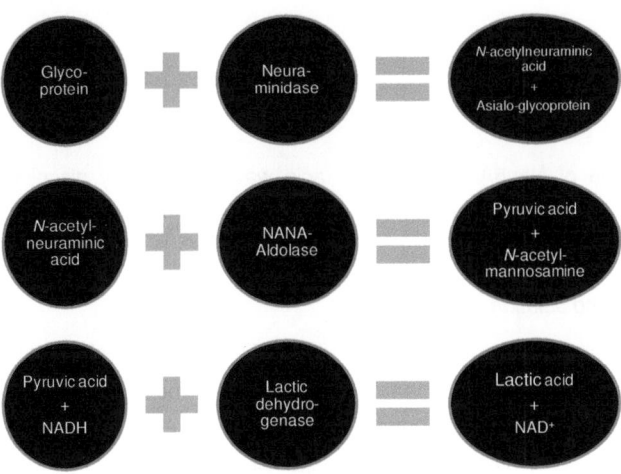

Figure 1: Enzymatic reaction schema

7.2 Material and Methods

7.2.1 Chemicals

All chemical used were of analytical grade. *N*-acetylneuraminic acid (Neu5Ac) and NADH were purchased from Calbiochem (USA). NANA-aldolase (ED 4.1.3.3, from *E. coli*), lactic dehydrogenase (EC 1.1.1.27, from bovine heart), glycine and Tris were purchased from Fluka (Switzerland). Neuraminidase (ED 3.2.1.18, from *Athrobacter ureafaciens*) was purchased from Roche Diagnostics (Germany). Pyruvic acid was purchased from Sigma-Aldrich (Germany). Sodium acetate, sodium cholate and 25% HCl were purchased from Merck (Germany). The fusion protein rhEpo-Fc, consisting of two molecules recombinant human erythropoietin attached to the Fc part of a human IgG_1 molecule, was expressed in Chinese Hamster Ovary cells (DUKK-B11, ATCC CRL-9096) and purified by affinity chromatography on Protein A Sepharose FF (GE Healthcare, Sweden).

7.2.2 Equipment

The measurement was carried out with a Microplate Spectrofluorometer (SPECTRAmax GEMINI XS, Molecular Devices, USA). The Validata 3.02.52 software [P.A.ASA (Arbeitsgruppe für Spurenanalyse), Graz, Austria] was used to determine all validation results.

7.2.3 Standard solutions

7.2.3.1 Stock solution

A stock solution of Neu5Ac and pyruvic acid was prepared by dissolving 50 mg and 30 mg, respectively in water to obtain a concentration of 10 µmol/ml. The solution was stored at 4°C.

7.2.3.2 Working solution

Each working solution was prepared up daily by diluting each stock solution with water to obtain a concentration of 100 nmol/ml. From this working solutions, aliquots with concentrations of 2, 5, 7.5 and 10 nmol/ml were made in sample buffer II (0.1 M glycine, 0.25 M sodium acetate, 0.03 M sodium cholate, pH 5.0).

7.2.4 Enzymatic assay

All samples were first prediluted in sample buffer I (0.1 M glycine, 0.05 M sodium acetate, pH 5.0) followed by 1:2 dilution in sample buffer II, so that each sample contained 0.015 M sodium cholate. 80 µl of the sample with a concentration between 4–20 nmol/ml, 80 µl sample buffer II and 40 µl neuraminidase (0.5 U/ml) were added into an eppendorf tube and incubated at 37°C for 18 ±2h. After incubation, 400 µl of 0.1 M Tris/HCl pH 8.2, 200 µl of NANA-aldolase solution (9.7 U/ml) and 200 µl of NADH solution (15 µM) were added and incubated at 37°C for 20 min. Thereafter, 200 µl of the reaction mixtures (fourfold) were added into a polystyrene 96-well microtiterplate (Fluotrac 200, Greiner, Germany). The fluorescence signals of these mixtures were measured at 30°C, using an excitation wavelength of 340 nm and an emission wavelength of 465 nm, for at least 3 times (0, 15 and 30 min). The reaction was started by adding 50 µl of the enzyme lactic dehydrogenase (26.6 U/ml). The 100% NADH solution was prepared similarly, but with 200 µl sample buffer II instead of the sample and neuraminidase. To calculate the calibration function of Neu5Ac working standard, the measured rfu values of the Neu5Ac working standard (15 and 30 min) were subtracted from the rfu values of the 100% NADH solution. All sample concentrations were calculated with this working standard equation.

To verify the Neu5Ac cleavage efficiency and the transformation to pyruvic acid by NANA-aldolase, we analyzed the Neu5Ac content of a defined glycoprotein stock solution with a known Neu5Ac content. This internal control sample (ICS) was analyzed routinely at each plate (Table 1).

Table 1: Schema of the microtiterplate

100% NADH	100% NADH	Neu5Ac 10 nmol/ml	Neu5Ac 5 nmol/ml	Pyruvic acid 10 nmol/ml	Pyruvic acid 5 nmol/ml	Sample 1	Sample 3	Sample 5	Sample 7	Sample 9	Internal control sample
100% NADH	100% NADH	Neu5Ac 7.5 nmol/ml	Neu5Ac 2 nmol/ml	Pyruvic acid 7.5 nmol/ml	Pyruvic acid 2 nmol/ml	Sample 2	Sample 4	Sample 6	Sample 8	Sample 10	Internal control sample

7.3 Validation parameter

The enzymatic microtiterplate assay was validated according to the ICH guidelines for validation of analytical procedures. For the calibration, N-acetylneuraminic acid working solution (Neu5Ac) and a representative purified rhEpo-Fc fusion protein sample was used as an internal control sample (ICS).

7.3.1 Specificity

To verify the specificity of the enzymatic assay, a control sample was analyzed according to the method described above followed by analysis without neuraminidase and NANA-aldolase. By this procedure, unspecific background levels can be evaluated. To prove the presence of free Neu5Ac in the sample solution, the enzymatic assay was performed without neuraminidase and NANA-aldolase.

The fluorescence signal measurement was carried out using a microtiterplate spectrofluorometer (SPECTRAmax GEMINI XS, Molecular Devices, USA) at an excitation wavelength of 340 nm and an emission wavelength of 465 nm.

7.3.2 Linearity, range and variance homogeneity

The calibration curve was obtained with four concentrations of the Neu5Ac working solution (2 – 10 nmol/ml) in sixfold analysis. The results were calculated with the program "Validata". On the basis of this analysis, the linearity, range and variance homogeneity were evaluated statistically.

7.3.3 Limit of detection and quantitation

The parameters limit of detection (LOD) and quantitation (LOQ) were calculated on the basis of the standard deviation (SD) of the response and the slope of the calibration curve at the lowest concentration.

$$LOD = \frac{3.3 \times \sigma}{S} \qquad LOQ = \frac{10 \times \sigma}{S}$$

σ = SD of the response
S = Slope of the calibration curve

7.3.4 Accuracy

In the absence of a commercial available standard, a purified rhEpo-Fc sample (ICS) was used. The accuracy was determined by recovery of known amounts of Neu5Ac working solution (5 and 10 nmol/ml) added to the ICS. All solutions were prepared in triplicate. Additionally, to obtain the relative accuracy, the Neu5Ac concentration of the ICS was analyzed during 23 independent analyses. On the basis of these analyses, the mean value, SD and the coefficient of variation (CV%) were calculated.

7.3.5 Precision

The precision of the enzymatic assay was determined by repeatability (intra-day), intermediate precision (inter-day and operator) and reproducibility. Repeatability was evaluated by assaying the ICS ten times at one day. The intermediate precision was determined by comparing the ICS results of two operators, which were obtained at two different days. All measurements were performed in triplicate. To evaluate the reproducibility we analyzed the ICS in six independent experiments, whereas each experiment was analyzed in triplicate. On the basis of these analyses, the mean value, SD and CV% were calculated.

7.3.6 Robustness

The robustness was verified by analyzing the Neu5Ac working solution (10 nmol/ml) in different microtiterplates, neuraminidase cleavage was verified at different incubation intervals followed by the fluctuation testing of the rfu values of the 100% NADH solution.

7.4 Appraisal factors

The validation parameters, listed in paragraph 7.3, were evaluated statistically. Whereas, only data which have analogy with the following acceptance criteria were used.

7.4.1 Acceptance criteria

- The Neu5Ac and pyruvic acid calibration curve must be linear in the range of 2 – 10 nmol/ml.
- The calibration function must be in the range of y = a + b * x [a = 0-6 and b = 4–7].
- The correlation value R^2 must be ≥ 99%.

- The rfu value of the 100% NADH solution must be above 80 rfu.
- The coefficient of variation (CV%) of the fourfold determination must be ≤ 5.
- At least two of four measurement points of each fourfold analyzed sample must be valid; otherwise the sample must be repeated.

7.5 Results

In the first step, the objective was to prove method specificity and linearity to evaluate the quantification range and the limits of detection (LOD) and quantitation (LOQ).

7.5.1 Specificity

As shown in Table 2, a control sample was analyzed according to the method described above, followed by analysis without neuraminidase and NANA-aldolase.

Table 2: Specificity

Experiment	Enzyme		Concentration [nmol/ml]
	Neuraminidase	NANA-aldolase	
1	+	+	19.31
2	-	+	nd[a]
3	-	-	nd[a]

[a] nd, not detectable

Without neuraminidase and NANA-aldolase treatment, no Neu5Ac was detected, thus this method was specific for the detection of Neu5Ac from purified samples. Neither the enzyme solution, nor the buffer substances influenced this enzymatic assay.

7.5.2 Linearity

The calibration curve was obtained with four concentrations of the Neu5Ac working solution (2 – 10 nmol/ml) in sixfold analysis. The results were calculated with the program "Validata". On the basis of these analyses, the linearity was evaluated statistically (Table 3).

Table 3: Linearity

CODE	Concentration [nmol/ml]	Δ rfu	CODE	Concentration [nmol/ml]	Δ rfu
SIA 10	10	61.47	SIA 7.5	7.5	46.96
SIA 10	10	61.16	SIA 7.5	7.5	46.11
SIA 10	10	62.38	SIA 7.5	7.5	47.65
SIA 10	10	61.15	SIA 7.5	7.5	43.71
SIA 10	10	61.16	SIA 7.5	7.5	44.16
SIA 10	10	60.98	SIA 7.5	7.5	43.83
	Mean value	61.39		Mean value	45.40

CODE	Concentration [nmol/ml]	Δ rfu	CODE	Concentration [nmol/ml]	Δ rfu
SIA 5	5	30.58	SIA 2	2	15.32
SIA 5	5	29.60	SIA 2	2	14.02
SIA 5	5	30.63	SIA 2	2	13.48
SIA 5	5	30.13	SIA 2	2	13.18
SIA 5	5	32.02	SIA 2	2	14.29
SIA 5	5	31.65	SIA 2	2	12.70
	Mean value	30.72		Mean value	13.83

Table 3 shows the Δrfu results of 4 different Neu5Ac concentrations in the range of 2-10 nmol/ml and the resulting mean values.

<u>Linearity:</u>   Test value   5.26
              F-value (99% level)   8.02

These results demonstrated that on the 99% level, no significant difference could be detected (Fig. 2). On the basis of these analyses, range and variance homogeneity were evaluated statistically.

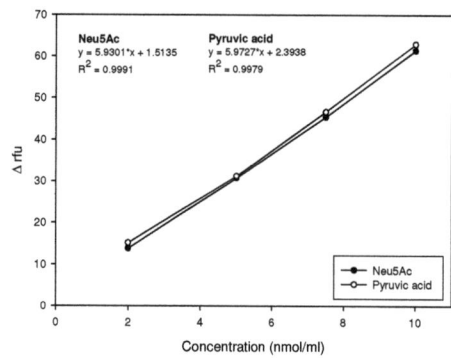

Figure 2: Calibration function and correlation value R^2 of the Neu5Ac and pyruvic acid working solution in the range of 2 – 10 nmol/ml.

Related to the calibration function, no differences were detected.

7.5.3 Range

On the basis of the data listed in Table 3, the range was calculated statistically.

Result uncertainty (k)	3	
Estimated measured value (0)	1.51	Δ rfu
Repetitions (samples)	6	
Decision level NWG	0.95	
t-value (1/2 sided)	1.72–2.07	
Decision level VB	0.95	
Factors VB	0.77–1.42	
Critical value	2.79	Δ rfu
Detection limit	0.21	nmol/ml
VB detection limit	0.17–0.30	nmol/ml
Acquisition limit	0.43	nmol/ml
VB acquisition limit	0.33–0.61	nmol/ml
Determination limit	0.74	nmol/ml
VB determination limit	0.58–1.05	nmol/ml

These results demonstrated that the quantification range is valid.

7.6 Variance homogeneity

On the basis of the data listed in Table 3, the variance homogeneity was calculated statistically.

Variance analysis:
Variance of the highest concentration	0.26
Variance of the lowest concentration	0.86
Test value (variance homogeneity)	3.25
F-value (95% level)	5.05
F-value (99% level)	10.97

These data demonstrated variance homogeneity, thus the test value was lower than the F-value at both levels of significance.

7.6.1 Limit of detection and quantitation

The limit of detection (LOD) and the limit of quantitation (LOQ) were calculated on the basis of the Δ rfu values of the lowest concentration.

$$LOD = \frac{3.3 \times \sigma}{S} = \frac{3.3 \times 0.93}{5.93} = 0.52 \ \ nmol/ml$$

$$LOQ = \frac{10 \times \sigma}{S} = \frac{10 \times 0.93}{5.93} = 1.57 \ \ nmol/ml$$

The LOD and LOQ amounted 0.52 nmol/ml, respectively 1.57 nmol/ml. Thus, the lowest quantification level of 2 nmol/ml was verified.

7.6.2 Accuracy

No commercial standard was available, thus a purified rhEpo-Fc sample was defined as an internal control sample (ICS).

7.6.2.1 Relative accuracy

The relative accuracy was calculated on the basis of 23 independent analyses.

Table 4: Relative accuracy

Experiment	Concentration [nmol/ml]		Statistic
1	14.72	Mean [nmol/ml]	14.46
2	13.57	SD [nmol/ml]	0.64
3	13.51	CV [%]	4.43
4	13.74		
5	14.22		
6	14.68		
7	14.95		
8	14.22		
9	15.61		
10	15.00		
11	14.69		
12	13.51		
13	14.03		
14	14.28		
15	13.90		
16	14.39		
17	14.02		
18	14.94		
19	14.85		
20	13.83		
21	15.46		
22	15.35		
23	15.15		

The relative accuracy of 23 independent analyses of the ICS was 14.46 nmol/ml by a SD of ± 0.64 nmol/ml and a CV% of 4.43 (Table 4).

7.6.2.2 Spike experiment

The recovery was verified by spiking various known amounts of Neu5Ac working solution (5 and 10 nmol/ml) into the ICS. All solutions were prepared in triplicate.

Table 5: Accuracy and recovery of Neu5Ac spike experiments

Experiment	Calculated concentration [nmol/ml]	Measured concentration [nmol/ml]	Recovery [%]	Mean	SD	CV%
1		17.60	90.46			
2		17.47	89.79			
3	19.46	18.97	97.46	91.86	0.56	3.14
4		17.86	91.76			
5		17.50	89.94			
6		17.85	91.74			
1		22.43	91.70			
2		22.21	90.80			
3	24.46	22.44	91.73	90.64	0.43	1.97
4		22.64	92.58			
5		21.83	89.24			
6		21.48	87.80			

The data presented in Table 5 showed recoveries between 87.80 – 97.46% at all levels.

7.6.3 Precision

7.6.3.1 Repeatability

Repeatability was evaluated by a tenfold determination of the ICS performed at one day.

Table 6: Repeatability

Experiment	Concentration [nmol/ml]	Statistic	
1	14.02	Mean [nmol/ml]	14.74
2	14.41	SD [nmol/ml]	0.51
3	14.87	CV [%]	3.48
4	15.11		
5	14.93		
6	15.70		
7	14.19		
8	14.83		
9	15.08		
10	14.29		

The mean concentration of the ICS was 14.74±1.94% nmol/ml to the relative accuracy. The SD and the CV% of the repeatability was 0.51 nmol/ml and 3.48, respectively (Table 6).

7.6.3.2 Reproducibility

To evaluate the reproducibility we were analyzing the ICS in six independent experiments.

Table 7: Reproducibility

Experiment	Concentration [nmol/ml]		
1	15.32	Mean [nmol/ml]	15.14
	14.92	SD [nmol/ml]	0.21
	15.18	CV [%]	1.36
2	15.52	Mean [nmol/ml]	15.52
	15.81	SD [nmol/ml]	0.29
	15.23	CV [%]	1.87
3	14.20	Mean [nmol/ml]	14.10
	14.18	SD [nmol/ml]	0.15
	13.93	CV [%]	1.10
4	15.17	Mean [nmol/ml]	15.55
	15.71	SD [nmol/ml]	0.33
	15.76	CV [%]	2.10
5	15.39	Mean [nmol/ml]	15.43
	15.60	SD [nmol/ml]	0.15
	15.32	CV [%]	0.96
6	15.48	Mean [nmol/ml]	15.43
	15.03	SD [nmol/ml]	0.38
	15.78	CV [%]	2.46

The mean values ranged between 97.51 – 107.53% compared to the relative accuracy. The SD ranged between 0.15 – 0.38 nmol/ml and the CV% between 0.96 – 2.46 (Table 7). Based on this validation parameter, samples could be analyzed within a 10% range under reproducible conditions.

7.6.3.3 Intermediate precision

To verify the intermediate precision, the ICS was analyzed by two operators at two different days.

Table 8: Intermediate precision

	Operator 1		Operator 2	
	Day 1	Day 2	Day 1	Day 2
Concentration [nmol/ml]	14.74[a]	15.31[a]	14.55[a]	15.14[a]
Mean [nmol/ml]	15.02		14.84	
SD [nmol/ml]	0.39		0.35	
CV%	2.63		2.36	

[a] (n=3)

The variance between the two operators was 1.2%, meaning that both operators could analyze the sample with similar precision. The SD and the CV% were in the same range.

7.6.4 Robustness

7.6.4.1 Microtiterplate charge

The robustness was verified by analyzing the Neu5Ac working solution (10 nmol/ml) in three different microtiterplate batches.

Table 9: Microtiterplate batches

CODE	Variation	Concentration [nmol/ml]	Mean [nmol/ml]	Recovery [%]
SIA 10	Batch 1	10.09 10.06	10.08	100.76
	Batch 2	10.14 10.22	10.25	102.52
	Batch 3	10.06 10.11	10.09	100.86

The recovery ranged between 100.76 – 102.52% compared to the working solution (Table 9). Thus, the enzymatic assay is not influenced by the microtiterplate batch.

7.6.4.2 Neuraminidase digestion

The enzymatic digestion of Neu5Ac by neuraminidase was verified by changing the incubation time.

Table 10: Neuraminidase digestion

Experiment	Incubation time [h]	Concentration [nmol/ml]		Statistic	Recovery [%]
1	16	14.27	Mean [nmol/ml]	14.35	99.25
		14.35	SD [nmol/ml]	0.08	
		14.43	CV [%]	0.55	
2	17	15.27	Mean [nmol/ml]	15.05	104.06
		15.55	SD [nmol/ml]	0.64	
		14.33	CV [%]	4.24	
3	18	14.55	Mean [nmol/ml]	14.74	101.95
		15.03	SD [nmol/ml]	0.25	
		14.65	CV [%]	1.70	
4	19	15.02	Mean [nmol/ml]	15.31	105.85
		15.29	SD [nmol/ml]	0.30	
		15.61	CV [%]	1.94	

The SD and the CV% ranged between 0.08 – 0,64 and 0.55 – 4.24, respectively (Table 10). No temporal trend was observed.

7.6.4.3 Influence of the rfu value of the 100% NADH solution

Table 11: Fluctuation testing of the rfu value of the 100% NADH solution

Solution	rfu value	Concentration [nmol/ml]		Statistic	Recovery [%]
1	103.87	10.10	Mean [nmol/ml]	10.09	100.90
		10.06	SD [nmol/ml]	0.03	
		10.12	CV [%]	0.27	
2	96.94	10.06	Mean [nmol/ml]	10.10	101.00
		10.14	SD [nmol/ml]	0.04	
		10.11	CV [%]	0.39	
3	90.96	10.17	Mean [nmol/ml]	10.23	102.30
		10.25	SD [nmol/ml]	0.05	
		10.26	CV [%]	0.46	

The fluctuation of the rfu values from the defined 100% NADH solution varied between 90.96 and 103.87 (Table 11). One working solution (10 nmol/ml) was measured with this varying rfu values, resulting in recoveries between 100.90 – 102.30% with SD between 0.03 – 0.05 nmol/ml and CV% between 0.27 – 0.46. This experiment demonstrates that varying NADH rfu values have no influence on the final result.

7.7 Discussion

To verify the specificity of the method, a control sample was analyzed according to the method described above and by changing the enzyme addition. Without the addition of neuraminidase and NANA-aldolase, no Neu5Ac was detected. For this reason, this method proved to be specific for the detection of Neu5Ac from purified protein samples. Neither the enzyme solution, nor the buffer impurities had an influence on the enzymatic method.

First of all, the linearity, the quantification range and the variance homogeneity were tested statistically. Each standard concentration was analyzed six times. The results were calculated with the statistic program Validata 3.02.52 software. Based on these analyses, linearity, quantification range and variance homogeneity could be demonstrated. Afterwards, the limit of detection (LOD) and the limit of quantitation (LOQ) were calculated on the basis of the Δ rfu values at the lowest concentration. The limit of detection was 0.52 nmol/ml and the limit of quantitation was 1.57 nmol/ml, thus the lowest quantification level of 2 nmol/ml was verified.

Furthermore, accuracy, precision and robustness were tested. Because of the absence of a commercial available standard, a purified rhEpo-Fc sample (internal control sample - ICS) was used. The relative accuracy of 23 independent analyses was 14.46 nmol/ml, with a standard deviation of 0.64 nmol/ml and a CV% of 4.43. The recovery was verified by spiking this sample with various amounts of Neu5Ac working solutions. The recovery ranged between 87.80 – 97.46% with a standard deviation between 0.13 – 0.83 nmol/ml. The CV % ranged between 0.58 – 4.59%.

The repeatability was verified with a standard deviation of 0.51 nmol/ml, a CV% of 3.48 and a deviation to the relative accuracy of ±1.94%. The standard deviation and the CV% of the accuracy and the repeatability were similar (SD_A = 0.64 nmol/ml, SD_R = 0.51 nmol/ml, $CV\%_A$ = 4.43 and $CV\%_R$ = 3.48).

To evaluate the reproducibility, the ICS was analyzed in six independent experiments, whereas each experiment was analyzed three times. The reproducibility ranged between 97.51 – 107.53. The standard deviation ranged between 0.15 – 0.38 nmol/ml and the CV% between 0.96 – 2.46.

These results demonstrate that samples can be analyzed within ± 10%.

To verify the intermediate precision, the ICS was analyzed by two operators at two different days. The variance between the two operators was 1.2%, whereas no significant differences regarding the SD and CV% were obvious.

The robustness was verified by analyzing the Neu5Ac working solution (10 nmol/ml) in different microtiterplate batches, the neuraminidase cleavage was varied at different incubation intervals and the fluctuation of the Δ rfu values of the NADH concentration in the 100% NADH solution.

The Neu5Ac working solution was analyzed in three different microtiterplate batches. The recovery ranged between 100.76 – 102.52%.

The enzymatic digestion of Neu5Ac by neuraminidase was verified by changing the incubation time (18±2 h). The standard deviation and the CV% were in the range of the precision. No temporal trend was seen.

The fluctuation of the rfu values of the 100% NADH solution varied between 90.97 and 103.87. Based on the recoveries (100.90 – 102.30%) of the working solution, the fluctuation of the rfu values has no influence on the final result.

In conclusion, the enzymatic assay for the determination of the Neu5Ac in the range of 2 – 10 nmol/ml is quantifiable with a first order equation. The validation documents showed that this micro method is robust, accurate and reproducible for the intended purpose. In summary, the proposed method is suitable for the analysis of purified glycoprotein samples for instance for routine quality control as well as clone screening.

7.8 References

[1] ICH-Guideline: Validation of Analytical Procedures: Methodology (6. November 1996) ICH Topic Q 2 A: Validation of Analytical Methods: Definitions and Terminology (June 1995)

[2] ICH Topic Q 2 B: Validation of Analytical Methods: Methodology (June 1997)

7.9 Formulary

Mean
$$\text{Mean} = \frac{\sum y_i}{n_i}$$

Standard Deviation (SD)
$$SD = \sqrt{\frac{\sum(y_i - \bar{y})^2}{n-1}}$$

Relative variation coefficient (CV%)
$$CV\% = \frac{100 \times s}{\text{Mean}}$$

Recovery (%)
$$R[\%] = \frac{100 \times \text{Mean}}{\text{Measured Concentration}}$$

Variance (%)
$$S[\%] = 100 - \left[\left(\frac{100 \times \text{Mean Operator1}}{\text{Mean Operator2}}\right)\right]$$

Limit of quantitation (LOQ)
$$LOQ = \frac{10 \times SD}{\text{slope}}$$

Limit of detection (LOD)
$$LOD = \frac{3.3 \times SD}{\text{slope}}$$

8 Publications

Biochemical characterization of rhEpo-Fc fusion protein expressed in CHO cells
Kornelia Schriebl, Evelyn Trummer, Christine Lattenmayer, Robert Weik, Renate Kunert, Dethardt Müller, Hermann Katinger, Karola Vorauer-Uhl
Protein Expression and Purification 49 (2006) 265 – 275

A novel strategy for quantitative isoform detection directly performed from culture supernatant
Kornelia Schriebl, Evelyn Trummer, Robert Weik, Dethardt Müller, Renate Kunert, Christine Lattenmayer, Hermann Katinger, Karola Vorauer-Uhl
Journal of Pharmaceutical and Biomedical Analysis 42 (2006) 322 – 327

Applicability of different fluorescent dyes for isoform quantification on linear IPG gels
Kornelia Schriebl, Evelyn Trummer, Robert Weik, Christine Lattenmayer, Dethardt Müller, Renate Kunert, Hermann Katinger, Karola Vorauer-Uhl
Electrophoresis 28 (2007) 2100 – 2107

Due to legal formality this paper cannot be printed.

Biochemical characterization of rhEpo-Fc fusion protein expressed in CHO cells

Kornelia Schriebl [a,*], Evelyn Trummer [a,b], Christine Lattenmayer [a,b], Robert Weik [c], Renate Kunert [b], Dethardt Müller [b], Hermann Katinger [b,c], Karola Vorauer-Uhl [b]

[a] *Austrian Center of Biopharmaceutical Technology, Muthgasse 18, A-1190 Vienna, Austria*
[b] *Institute of Applied Microbiology, Department of Biotechnology, University of Natural Resources and Applied Life Science, Muthgasse 18, A-1190 Vienna, Austria*
[c] *Polymun Scientific GmbH, Nussdorferlaende 11, A-1190 Vienna, Austria*

Received 9 March 2006, and in revised form 28 April 2006
Available online 15 June 2006

Abstract

One challenge in biotechnology industry is to produce recombinant proteins with prolonged serum half-life. One strategy for enhancing the serum half-life of proteins includes increasing the molecular weight of the protein of interest by fusion to the Fc part of an antibody. In this context, we have expressed a homodimer fusion protein in CHO cells which consists of two identical polypeptide chains, in which our target protein, recombinant human erythropoietin (rhEpo), is N-terminally linked with the Fc part of a human IgG$_1$ molecule. In the present study, culture supernatant of a stable clone was collected and purified by affinity chromatography prior characterization. We emphasized product quality aspects regarding the fusion protein itself and in addition, post-translational characterization of the subunits in comparison to human antibodies and rhEpo. However, overproduction of recombinant proteins in mammalian cells is well established, analysis of product quality of complex products for different purposes, such as product specification, purification issues, batch to batch consistency and therapeutical consequences, is required. Besides product quantification by ELISA, N-acetylneuraminic acid quantification in microtiterplates, quantitative isoform pattern and entire glycan profiling was performed. By using these techniques for the characterization of the recombinant human Epo-Fc (rhEpo-Fc) molecule itself and furthermore, for the separate characterization of both subunits, we could clearly show that no significant differences in the core glycan structures compared to rhEpo and human antibody N-glycans were found. The direct comparison with other rhEpo-Fc fusion proteins failed, because no appropriate data were found in the literature.
© 2006 Elsevier Inc. All rights reserved.

Keywords: CHO cells; rhEpo-Fc fusion protein; N-Glycan; N-Acetylneuraminic acid; Quantitative isoform pattern

---

Fc fusion proteins are promising dimers in respect of new potent drug design. In this regard, several fusion proteins containing a target protein which comprises, from its N-terminal to C-terminal direction, an immunoglobulin Fc region which lacks at least the CH1 domain, in certain cases a peptide linker and an amino acid sequence of the target protein. First patents, concerning Fc fusion protein technology were published in the late nineties [1]. The motivation to create non natural occurring proteins is to compensate therapeutical drawbacks such as short biological half-life or degradation processes. Beside prolonged serum half-life, alternative drug delivery applications such as pulmonary application are developed [2]. For fusion partners Fc domains from IgG$_1$, IgG$_2$, IgG$_3$ and IgG$_4$ are used. As target proteins primarily glycoproteins such as cytokines [1] and different hormones [3] were fused. As one example of a highly glycosylated polypeptide, the human

---

[*] Corresponding author. Fax: +43 1 3697315.
E-mail address: kschriebl@gmx.at (K. Schriebl).

1046-5928/$ - see front matter © 2006 Elsevier Inc. All rights reserved.
doi:10.1016/j.pep.2006.05.018

erythropoietin (Epo),[1] which is a 30.4 kDa glycoprotein hormone produced by the kidney in adult humans is described [4,5]. It consists of a 165 amino acid single polypeptide chain containing two disulphide bonds [6,7] and has three N-linked (Asn-24, Asn-38, Asn-83) and one O-linked (Ser-126) sugar chain [8,9]. Its microheterogeneity is related to the charged carbohydrate moiety of the protein and was studied extensively [10,11]. The microheterogeneity of Epo is seen on the N-linked carbohydrate chains, where the oligosaccharide may contain bi-, tri- and tetraantennae, each of which is typically terminated with the negatively charged sialic acid molecule. In human application, recombinant human erythropoietin (rhEpo) results in poor pharmacokinetics and the redesign of a fusion protein to promote a longer serum half-life is an important medical and commercial goal.

For the recombinant production of Epo-Fc fusion proteins different mammalian host cell lines such as CHO, NS/0, PerC6 and BHK are described [12]. Products isolated from these expression systems were primarily characterized in respect of their pharmacokinetic profiles and potency.

But, beside the medical aspects of recombinant human Epo-Fc (rhEpo-Fc) fusion protein biochemical analysis is important. On the one hand the similarity of the fusion protein to the natural occurring partners and on the other hand host cell line depending differences may be taken into account. For this reason, we have chosen the rhEpo-Fc fusion protein as a model protein to establish a technology platform for the development of new cell technology, cell physiology studies and analytical aspects. We have expressed a homodimer fusion protein in CHO cells which consists of two identical polypeptide chains, in which rhEpo is N-terminally linked with the Fc part of a human IgG$_1$ molecule. The fusion protein rhEpo-Fc comprises of 798 amino acids and has a molecular weight of about 112 kDa, of which about 89 kDa is contributed by the polypeptide backbone. This fusion protein represents a very heterogeneous, complex glycan and isoform pattern, whereas the isoelectric migration is primarily determined by the sialylated glycans of the Epo domains. The aim of this study was to characterize the biochemical properties of the rhEpo-Fc fusion protein. RhEpo-Fc characteristics were compared with well-known characteristics of the rhEpo and human IgG, because of the lack of quantitative information of other rhEpo-Fc molecules. N-Glycan analysis of the rhEpo-Fc molecule as well as of the rhEpo and recombinant human IgG$_1$ Fc (rhFc) subunit was performed to assign the N-glycan structure to the appropriate molecule. Furthermore, the N-acetylneuraminic acid content and the quantitative isoform pattern were analyzed.

Materials and methods

Cell line, culture medium and cultivation

Dihydrofolate reductase-deficient Chinese Hamster Ovary cells (DUKX-B11, ATCC CRL-9096) were co-transfected with genes for fusion protein rhEpo-Fc and dihydrofolate reductase (DHFR). The genetic construct consists of the Epo leader, the Epo coding sequence and the hinge-CH2-CH3 constant region of a human IgG$_1$ leading to a recombinant human Epo-Fc homodimer (Fig. 1) that is secreted into the culture supernatant. Transfected cells were selected for growth in the presence of 0.096 μM MTX. After revitalization from a research cell bank, cells were cultivated in suspension in Dulbecco's modified Eagle's medium DMEM/HAM'S F-12 (1:1 mixture), supplemented with 0.58 g/l L-glutamine, an in-house developed protein-free supplement (proprietary formulation), 0.25% soya peptone, 0.1% Pluronic F68 and 0.096 μM MTX. Within 2 weeks one selected CHO clone was adapted in spinner flasks (TECHNE, UK) to an in-house developed production medium (Polymun Scientific GmbH; Austria) of proprietary formulation. Cells were maintained in suspension culture in 125 and 500 ml spinner flasks on a magnetic stirrer plate (Techne, UK; Integra Biosciences, Switzerland) at 50 rpm and 37 °C (5% CO_2).

Purification of rhEpo-Fc

The secreted rhEpo-Fc from CHO cells was purified by affinity chromatography on Protein A Sepharose FF (GE Healthcare, Sweden). Cell-free culture supernatant samples containing rhEpo-Fc were adjusted to pH 8.5 with 1 M Tris, pH 9.0, and loaded onto a 314 μl HR 5 column at a flow rate of 0.2 ml/min (60 cm/h). After extensive washing of the column with 0.025 M Tris, 0.15 M NaCl, pH 8.5, rhEpo-Fc was eluted with 0.1 M glycine–HCl, pH 3.5 at a flow-rate adjusted to 30 cm/h. The eluted fraction was rapidly neutralized with 1 M Tris, pH 9.0. After elution, the column was regenerated with 0.1 M glycine–HCl, pH 2.5.

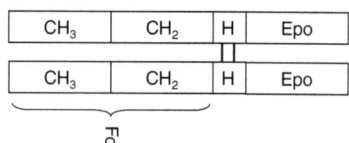

Fig. 1. Structure of N-terminally linked rhEpo with the Fc part of a human IgG$_1$ molecule.

---

[1] *Abbreviations used:* Epo, erythropoietin; rhEpo, recombinant human erythropoietin; DHFR, dihydrofolate reductase; SEC, size exclusion chromatography; ELISA, enzyme-linked immunosorbent assay; HRP, horseradish peroxidase conjugate; OPD, O-phenylenediamine dihydrochloride; IEF, Isoelectric focusing; CHAPS, 3-((3-cholamidopropyl)dimethylammonio)-1-propanesulfonate; BP, bandpass; LOQ, Limit of quantitation; ABA, Aminobenzoic acid; SD, standard deviation.

Papain digestion

For digestion 2 μl papain from *papaya latex* (10 μg/μl suspension, Sigma–Aldrich, Germany) was mixed with 18 μl freshly prepared activation buffer (0.05 M sodium phosphate buffer, 0.001 M EDTA, 0.01 M cysteine, pH 7.0) and activated by incubation at 37 °C for 10 min. Desalted, lyophilized rhEpo-Fc sample (2.2 nmol) was reconstituted into 200 μl digestion buffer (0.05 M sodium phosphate buffer, 0.001 M EDTA, pH 6.3). The cleavage was initiated by the addition of 12.5 μl papain solution to the rhEpo-Fc sample to give a papain/rhEpo-Fc ratio of 5% (w/w). This mixture was incubated at 37 °C for 30 min. Digestion was stopped by the addition of 6 μl of 1 M iodoacetamide.

Purification of the rhEpo and the rhFc subunit

After papain digestion, the sample was buffered into PBS by NAP-5 (GE Healthcare, Sweden). One milliliter sample was incubated with 100 μl Prosep-vA Ultra chromatography media (Millipore, Watford, UK) at room temperature on a shaker for 1 h. After incubation, the sample was centrifuged at 5000g for 0.5 min. The supernatant contained the rhEpo subunit, whereas the rhFc subunit was still binding to the Protein A. Prosep-vA media was washed three times with PBS and the rhFc subunit was eluted with 0.1 M glycine–HCl, pH 3.5. The eluted fraction was rapidly neutralized with 1 M Tris, pH 9.0. The rhEpo subunit as well as the rhFc subunit were desalted by PD-10 and NAP-5, respectively, and lyophilized. Papain digestion was monitored by size exclusion chromatography (SEC) on a TSKgel G3000SW column (TOSOH Biosciences, Stuttgart, Germany).

RhEpo-Fc quantification by ELISA

The rhEpo-Fc content at all steps was quantified with a high-sensitivity sandwich enzyme-linked immunosorbent assay (ELISA). Microtiterplates (MaxiSorb, Nunc) were coated with a goat anti human IgG (γ-chain specific) antibody (100 μl/well, 0.088 mg/ml, Sigma, Germany) in 0.1 M sodium bicarbonate buffer, pH 9.6 for 2 h at room temperature. For each assay, we performed a duplicate standard curve by utilizing serial dilutions (1:2) of rhEpo-Fc standard (2.23–0.01 pmol/ml) and rhEpo-Fc samples in PBS/0.1% Tween 20 (PBS-T) and 1% BSA to quantify sample concentrations. Fifty microliters per well of the standard and sample dilution series were applied and incubated for 1 h at room temperature. Thereafter, the plates were incubated with a goat anti human IgG (γ-chain specific) antibody horseradish peroxidase conjugate (HRP) (50 μl/well, 0.75 μg/ml, Zymed Laboratories, USA) in PBS-T/1% BSA for 1 h at room temperature followed by staining with O-phenylenediamine dihydrochloride (OPD, Fluka, Switzerland). Hundred microliters per well of staining solution (120 μl OPD (100 mg/ml), 12 ml (0.035 M citric acid monohydrate, 0.067 M sodium phosphate dibasic dihydrate, pH 5.0), 12 μl 35% hydrogen peroxide solution) were added and incubated at room temperature. The enzymatic reaction was stopped by applying 100 μl 25% sulphuric acid to each well. The optical density was read with an ELISA plate reader (Tecan Sunrise) at 492 nm with a reference wavelength of 620 nm. ELISA software Magellan was utilized for processing the standard curve and for calculation of the amount of rhEpo-Fc in the samples.

SDS–PAGE

The fusion protein rhEpo-Fc was analyzed by SDS–PAGE on a 4–20% Novex Tris–Glycine gel (Invitrogen, USA) under non-reducing and reducing conditions as well as after PNGase F treatment. Under non-reducing conditions the sample (1.8 nmol/ml) was diluted 1:2 in 2× sample buffer (Invitrogen, USA), whereas under reducing conditions the sample was treated with 5% 2-mercaptoethanol (2-ME, Fluka, Switzerland) at 100 °C for 10 min and diluted 1:2 in 2× sample buffer. For the deglycosylation under reduced conditions rhEpo-Fc (3.6 nmol/ml) was denatured at 100 °C for 10 min in 2% SDS and 10% 2-ME. PNGase F from *Flavobacterium meningosepticum* (10 U) (New England BioLabs, Hitchin, UK) digestion was carried out in 0.5 M sodium phosphate buffer, pH 7.5 with 4% (w/v) N-octanoyl-N-methylglucamine (MEGA-8, Fluka, Switzerland) for 20 ± 2 h at 37 °C. Proteins were stained by Coomassie brilliant blue [13] for 30 min followed by destaining with a solution containing 25% ethyl alcohol and 8% acetic acid.

Size exclusion chromatography

Size exclusion chromatography (SEC) was performed using an Agilent HP 1100 system with a VWD detector (Agilent Technologies, Palo Alto, USA). Operation and acquisition were performed using the ChemStation software package (Agilent Technologies, Palo Alto, USA). Separation of the rhEpo-Fc fusion protein was carried out on a TSK-Gel G3000SW column (7.5 × 600 mm, 10 μm; TOSHOH Biosciences, Stuttgart, Germany) with a TSK guard column SW (7.5 × 7.5 mm, 10 μm; TOSHOH Biosciences, Stuttgart, Germany). Purified rhEpo-Fc sample (50 μl; 1.1 nmol/ml) were eluted in a 80 min run at a flow rate of 0.5 ml/min. Detection was performed at 214 nm. The mobile phase consisted of 0.0015 M potassium dihydrogen phosphate, 0.008 M disodium hydrogen phosphate dihydrate, 0.4 M sodium chloride, pH 7.4, and was filtered through a 0.45 μm filter prior use.

Isoelectric focusing

Isoelectric focusing (IEF) was carried out on a linear IPG gel (pH 4–8) [14,15]. The rhEpo-Fc sample was treated with 2-D Clean Up Kit (GE Healthcare, Sweden) according to the kit instructions and resuspended in 0.03 M Tris,

3 M urea, 4% (w/v) 3-((3-cholamidopropyl)dimethylammonio)-1-propanesulfonate (CHAPS). Thereafter, 4 pmol/8 μl of the rhEpo-Fc sample was minimally labeled with 1 μl Cy3 (3.5 pmol/μl, GE Healthcare, Sweden) on ice for 30 min in the dark. The labeling reaction was quenched by the addition of 1 μl of 0.01 M lysine and subsequently incubated on ice for 10 min in the dark. Prior loading, Cy3 labeled sample was diluted with 4% (w/v) CHAPS to 20 μl and entirely loaded on a Multiphor II flat bed electrophoresis unit (GE Healthcare, Sweden) [15] with the following settings: 15 °C, 2.5 mA, 5 W, 0.075 kVh (0.75 h, 100 V, step-n-hold) 0.375 kVh (0.75 h, 500 V, step-n-hold) 0.500 kVh (0.5 h, 1000 V, step-n-hold) 72 kVh (18 h, 4000 V, step-n-hold). After focusing, the separated rhEpo-Fc isoforms were detected by scanning the gel using the Typhoon 9400™ variable mode imager (GE Healthcare, Sweden) with a 532 nm laser and an emission filter of 580 nm bandpass (BP) 30. Image analysis was carried out with the Image Quant TL v2003 software (GE Healthcare, Sweden).

Sialic acid determination

N-Acetylneuraminic acid (Neu5Ac) was determined using an enzymatic microtiterplate assay. Enzymatic cleavage was performed by neuraminidase from *Athrobacter ureafaciens*, followed by cleaving them to pyruvic acid and N-acetylmannosamine using N-acetylneuraminic acid aldolase. Pyruvic acid was assayed using lactic dehydrogenase coupled to the oxidizing of NADH. NADH oxidation to NAD^+ was quantified spectrofluorometrically. Briefly, all samples were first prediluted in sample buffer I (0.1 M glycine, 0.05 M sodium acetate, pH 5.0) followed by 1:2 dilution in sample buffer II (0.1 M glycine, 0.05 M sodium acetate, 0.03 M sodium cholate, pH 5.0) so that every sample contained 0.015 M sodium cholate. Eighty microliters of the sample with a concentration level between 4 and 20 nmol/ml Neu5Ac, 80 μl sample buffer II and 40 μl neuraminidase from *A. ureafaciens* (0.5 U/ml, Roche Diagnostics, Germany) were added into an Eppendorf tube and incubated at 37 °C for 18 ± 2 h. After the incubation, 400 μl of 0.1 M Tris–HCl, pH 8.2, 200 μl of NANA-aldolase solution (9.7 U/ml, Fluka, Switzerland) and 200 μl of NADH solution (15 μM, Calbiochem, USA) were added and incubated at 37 °C for 20 min. Thereafter, 200 μl of the reaction mixtures (fourfold) were added into a 96-well microtiterplate (Fluotrac 200, Greiner, Germany). The fluorescence signals of these mixtures were recorded at 30 °C (Ex 340 nm, Em 465 nm) using a Microplate Spectrofluorometer (SPECTRAmax GEMINI XS, Molecular Devices, USA). The reaction was started by adding 50 μl of the enzyme lactic dehydrogenase (26.6 U/ml, Fluka, Switzerland). Neu5Ac (Calbiochem, USA) is used as standard in a concentration range of 2–10 nmol/ml. Limit of quantitation (LOQ) for the assay amounts 1.57 nmol/ml. As an internal reference standard a defined glycoprotein standard is used and analyzed routinely at each plate.

Oligosaccharide determination

PNGase F digestion and purification of the released N-glycans

The release of N-glycans of desalted and lyophilized samples using PNGase F from *F. meningosepticum* was performed according to the kit instructions (New England BioLabs, Hitchin, UK). Briefly, lyophilized rhEpo-Fc (2.2 nmol), rhEpo and rhFc subunit (received of 2.2 nmol rhEpo-Fc by papain digestion) were dissolved in 64 μl water and denaturated with 6 μl denaturing buffer at 100 °C for 10 min. After cooling, 10 μl enzyme buffer (supplied as 10× concentration), 10 μl 10% NP-40 and 10 μl PNGase F (10,0000 U/ml) were added and incubated at 37 °C for 20 ± 2 h. After digestion, the denatured protein was precipitated by adding 75 μl water and 100 μl of 1 M acetic acid and removed by centrifugation at 5000g for 1.5 min. To maximize recovery of the released N-glycans, the protein precipitate was washed three times with 100 μl of 1 M acetic acid. All N-glycan containing fractions were collected and passed through a Oasis® HLB cartridge (Waters, Watford, UK) to remove soluble peptides and lyophilized in a v-vial.

Exoglycosidase enzyme digestion

Removal of terminal sialic acid from reduced rhEpo-Fc was performed with neuraminidase from *A. ureafaciens* (Roche, Germany). Neuraminidase cleaves terminal sialic acids linked via α 2–3 and 2–6 to galactose and α 2–8 to glycoproteins or glycolipids, whereas branched sialic acids are not affected. Neuraminidase digestion was performed in combination with the PNGase F digestion. Briefly, after the denaturation step and the addition of PNGase F, 10 μl of neuraminidase (10 U/ml) were added and incubated at 37 °C for 20 ± 2 h.

Release of terminal galactose from asialo rhEpo-Fc N-glycans was performed with β-galactosidase from *Streptococcus pneumoniae* (QA-Bio, San Mateo, USA). β-Galactosidase removes all non-reducing terminal galactose linked via β1–3,4,6 to N-acetylglucosamine, whereas branched galactose is not affected and fucose linked to the penultimate N-acetylglucosamine will block cleavage. Lyophilized asialo rhEpo-Fc N-glycans were dissolved in 13 μl water. After the addition of 4 μl of 0.25 M sodium phosphate buffer, pH 5.0, and 3 μl β-galactosidase (50 U/ml), the reaction mixture was incubated at 37 °C for 20 ± 2 h.

Release of terminal N-acetylglucosamine from asialo rhEpo-Fc N-glycans was performed with β-galactosidase and glucosaminidase from *S. pneumoniae* (QA-Bio, San Mateo, USA). Glucosaminidase removes all non-reducing terminal β-linked N-acetylglucosamine (GlcNAc), whereas bisecting GlcNAc slows the reaction. Lyophilized asialo rhEpo-Fc N-glycans were dissolved in 12 μl water. After the addition of 4 μl of 0.25 M sodium phosphate buffer, pH 5.0, 2 μl β-galactosidase (50 U/ml) and 2 μl glucosaminidase (1 U/ml), the reaction mixture was incubated at 37 °C for 20 ± 2 h.

After incubation, all reaction mixtures were lyophilized and labeled with 4-aminobenzoic acid (4-ABA).

Aminobenzoic acid (ABA) labeling

Lyophilized N-glycans (440 pmol) were reconstituted in 10 μl 2 M 4-ABA solution (freshly prepared in DMSO/acetic acid; 7:3) and 15 μl 1.5 M tetrabutylammonium cyanoborohydride solution (freshly prepared in DMSO/acetic acid; 7:3). After heating at 60 °C for 4 ± 0.1 h, 100 μl deionized water was added and excess reagent was removed by extraction three times with 300 μl ethyl acetate. The 4-ABA-labeled N-glycans, which were in the lower aqueous phase, were transferred into small v-vials and loaded onto the TSK-Gel Amide-80 column.

Normal-phase HPLC

N-Glycan analysis was performed using an Agilent HP 1100 system with a FLD detector (Agilent Technologies, Palo Alto, USA). Operation and acquisition were performed using the ChemStation software package (Agilent Technologies, Palo Alto, USA).

Separation of charged and neutral 4-ABA labeled N-glycans by normal-phase HPLC was carried out using a TSK-Gel Amide-80 column (4.6 × 250 mm, 5 μm; TOSOH Bioscience, Stuttgart, Germany). The flow rate was 1 ml/min and the column temperature was adjusted to 25 ± 2 °C. The injection volume was 10 μl. Detection was performed using an excitation wavelength of 296 nm and an emission wavelength of 359 nm. The mobile phase consisted of a gradient formed from solutions A (80% (v/v) acetonitrile, 20% 0.01 M ammonium acetate, pH 3.7, 0.0025 M trifluoroacetic acid) and solution B (20% (v/v) acetonitrile, 80% 0.01 M ammonium acetate, pH 3.7, 0.0025 M trifluoroacetic acid). The gradient was as follows: 25% B over 15 min, 25–41% B over the next 60 min, 41–25% B in 1 min, 25% B over the next 14 min.

N-Glycan standards were purchased from Dextra Laboratories (Reading, UK) and Glyko (Novato, CA). These comprised asialo agalacto biantennary (NGA2), asialo agalacto biantennary fucosylated (NGA2F), asialo biantennary (NA2), asialo biantennary fucosylated (NA2F), monosialo biantennary (A1), monosialo biantennary fucosylated (A1F), disialo biantennary (A2), disialo biantennary fucosylated (A2F), asialo triantennary (NA3), trisialo triantennary (A3), asialo tetraantennary fucosylated (NA4F) and tetrasialo tetraantennary fucosylated (A4) structures.

Results

Cultivation

Cells cultivated in a 500 ml spinner flask were harvested every 3–4 days with an average cell density of 8×10^5 cells/ml and a viability ⩾95%. The average volumetric titer, analyzed by ELISA, amounted 200 pmol/ml.

Purification

Purification was performed by affinity chromatography on Protein A Sepharose FF. On average, 96% of the loaded protein was recovered (Table 1). The Protein A purified rhEpo-Fc fusion protein was used for protein characterization without further purification.

Protein characteristics

The rhEpo-Fc fusion protein schematically shown in Fig. 1 represents a homodimer fusion protein which consists of two molecules of rhEpo attached to the Fc part of a recombinant human IgG_1 molecule including the hinge, CH2 and CH3 domains.

Determination of the molecular weight of the glycosylated and deglycosylated protein was performed by SDS–Page. Homogeneity of the fusion protein was determined by SDS–PAGE and size exclusion chromatography. Furthermore, SDS–PAGE was used to control the quality of the purified product. RhEpo-Fc containing samples were analyzed under reducing and non-reducing conditions. Due to the electrophoretic pattern, purified protein contains only homodimers (Fig. 2). Dimerization, respectively, aggregation, was determined by size exclusion chromatography and was in the range of 20%, whereas only samples in Tris–glycine buffer were analyzed. (Fig. 3). Purity of the protein according the SDS–PAGE was greater than 98%.

Table 1
Purification of rhEpo-Fc fusion protein

	Protein concentration (pmol/ml)	Total volume (ml)	Total protein (pmol)	Yield (%)
Crude protein	242	41.8	10116	100
Protein A Sepharose FF	6957	1.4	9740	96

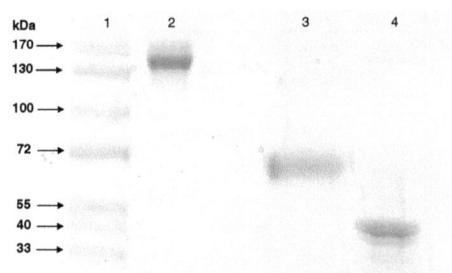

Fig. 2. Separation of rhEpo-Fc by SDS–PAGE (Tris–Glycine 4–20%) followed by Coomassie brilliant blue staining. Lane 1, molecular weight marker; lane 2, non-reduced rhEpo-Fc; lane 3, reduced rhEpo-Fc; lane 4, reduced deglycosylated rhEpo-Fc.

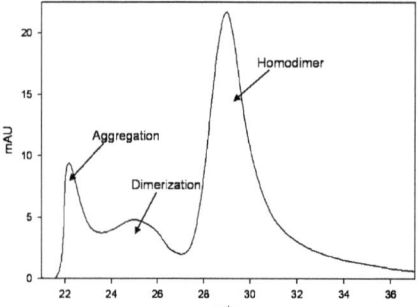

Fig. 3. Size exclusion chromatography (SEC) fractionation profiles of rhEpo-Fc. As the main peak the homodimer rhEpo-Fc was detected, followed by dimers and aggregates in the range of about 20%.

Isoform pattern

The determination of the individual isoform pattern is state of art in respect to product quality. By a more quantitative method, such as isoelectric focusing of prestained samples, the relative isoform distribution can be measured. The dimer configuration of the Fc fusion protein and the complex glycan pattern of the rhEpo fusion partner were responsible for 15 distinct isoforms in the pH range between 4.6 and 6.5, whereas the individual isoforms were randomly distributed in the range of 1 until 15% (Fig. 4).

Sialylation

The N-acetylneuraminic acid content of rhEpo-Fc was determined using an enzymatic microtiterplate assay. The

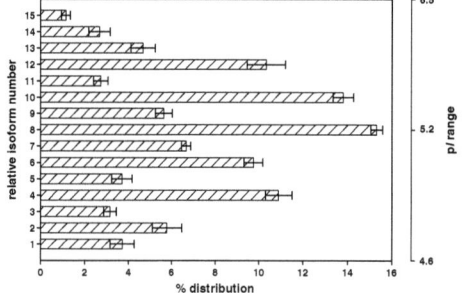

Fig. 4. Isoelectric focusing of a representative rhEpo-Fc sample on a linear IPG gel in the pH range 4–8. The rhEpo-Fc concentration was 4 pmol. The separated isoforms were labeled with Cy3 and scanned using the Typhoon 9400™ variable mode imager. The sum of all detectable isoforms were defined as 100%, followed by the calculation of the relative distribution for each individual isoform. The maximum SD of four independent analyses of several corresponding isoforms amounted ±0.8%.

concentration of various samples were in the range of 10–14 mol N-acetylneuraminic acid/mol rhEpo-Fc, with an average value of 12 mol N-acetylneuraminic acid/mol rhEpo-Fc with a standard deviation (SD) of ±0.48 nmol/ml and a CV percent of 3.94 ($n = 6$), which corresponds to a sialylation degree of 43%.

Oligosaccharide structure

To analyze N-glycan structures of rhEpo-Fc, normal-phase HPLC was performed. For this purpose, an optimized acetonitrile–ammonium acetate gradient was used performing the separation on an Amide-80 column. Most of the oligosaccharide standards were eluted in the form of one peak, while standards representing trisialylated triantennary (A3) and tetrasialylated tetraantennary fucosylated (A4) oligosaccharides eluted with three and four peaks. For this purpose, the area of the peaks which represented the A3 or A4 oligosaccharide were combined. As shown in Fig. 6, 16 different N-glycans could be identified. All oligosaccharide abbreviations are listed in Fig. 5. The predominant oligosaccharide structures of the rhEpo-Fc fusion protein were NA1F (13 %), A1F (15 %), A3 (15 %) and A4 (18 %). The maximum SD of four independent analyses of 16 different N-glycans amounted ±0.75%. About 60% of all complex N-glycans were fucosylated.

To analyze both subunits separately, lyophilized rhEpo-Fc was digested with papain followed by an appropriate purification procedure. To optimize cleavage times, papain digestion was performed in an interval from 10 to 60 min. Samples were analyzed every 10 min by size exclusion chromatography on a TSK-Gel G3000SW column (TOSOH Biosciences, Stuttgart, Germany), whereas optimized cleavage efficiency was determined after 30 min. N-Glycans of both subunits under optimized conditions were released by PNGase F digestion and analyzed in respect to their individual N-glycan structures.

As expected, the rhEpo as well as the rhFc subunit was glycosylated (Fig. 7). The predominant detected oligosaccharide structures of the rhFc subunit were NA1F (24%) and NA1 (20%) followed by NGA2F, NA2 and NA2F (each 12%). Approximately 50% of all N-glycans of the rhFc subunit were fucosylated. The predominant N-glycans of the rhEpo subunit consisted of A4 (34%), A3 (20%) and A1F (14%). All other neutral and charged structures were detected in small amounts, each <10%.

Sequential exoglycosidase digestion of rhEpo-Fc

Sequential glycan cleavage was performed to evaluate the structure analysis, as an alternative method to mass spectroscopy studies. After the stepwise removal of the corresponding monosaccharides, samples were analyzed. Results are shown in Fig. 8.

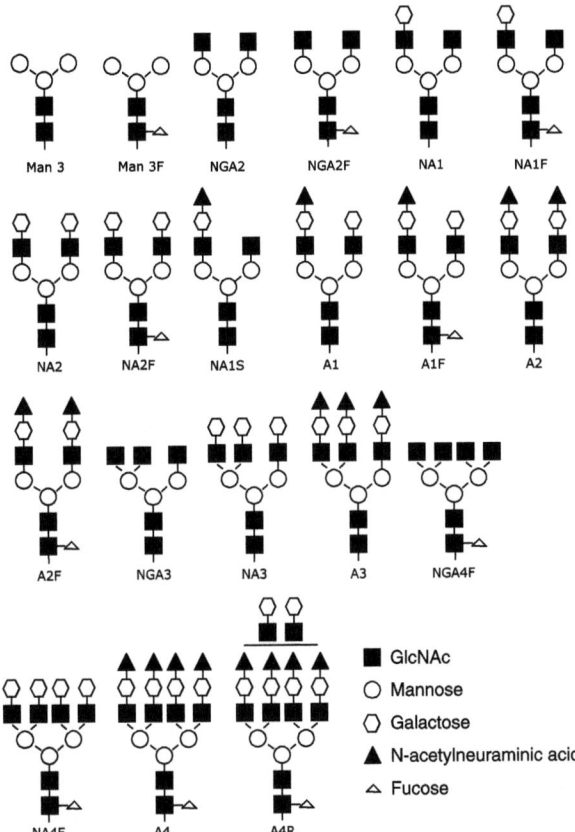

Fig. 5. Structures and names of oligosaccharide standards used of calibration purposes.

Discussion

One challenge in biotechnology industry is to produce recombinant protein with prolonged serum half-life. Thus, one strategy for enhancing the serum half-life of proteins includes increasing molecular weight by fusion of the protein to an Fc region of an antibody [16]. Gillies et al. have expressed rhEpo-Fc proteins in BHK cells [12]. These highly sialylated rhEpo-Fc fusion proteins had longer serum half-life and increased in vivo potency compared to recombinant human erythropoietin, to hyperglycosylated erythropoietin analogue NESP or to rhEpo-Fc fusion proteins of the same amino acid sequence expressed in NS/0, PerC6 or HEK-293 cells [12]. From our knowledge only poor information about the biochemical characterization is published, even though quantitative information should be given to specify this product entirely.

In the present study, we have characterized a homodimeric rhEpo-Fc fusion protein expressed in dihydrofolate reductase-deficient CHO cells. The fusion protein was isolated from the culture supernatant by Protein A affinity chromatography and was analyzed with respect to quantitative isoform pattern, sialylation, glycosylation and oligosaccharide structures.

As shown in Fig. 3, the fusion protein was mainly expressed in the dimeric configuration, but showed certain tendency to aggregate. This phenomenon was first described by Way et al. [17], who expressed a modified rhEpo-Fc fusion protein in NS/0 cells. According to his data, unmodified rhEpo-Fc tended to aggregate even more than modified fusion protein. Because this phenomenon was not detectable on SDS–PAGE it was suggested that folding characteristics were mainly responsible for aggregation. Dimerization as well as aggregation of single rhEpo

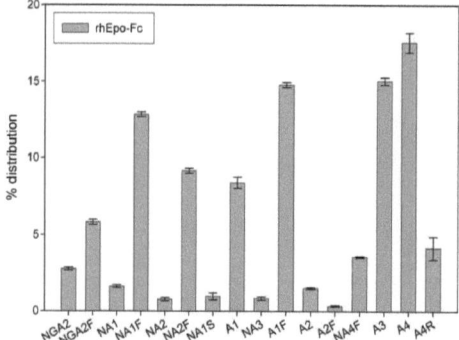

Fig. 6. The relative distribution was calculated from the normal-phase HPLC profile of 4-ABA labeled rhEpo-Fc N-glycans, the relative distribution was calculated. The rhEpo-Fc concentration was 2.2 nmol. The N-glycans were identified by comparison to oligosaccharide standards. The names of the different glycans are listed in Fig. 5.

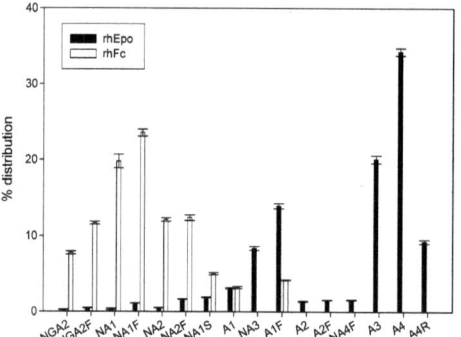

Fig. 7. The relative distribution was calculated from the normal-phase HPLC profiles of 4-ABA labeled rhEpo and rhFc N-glycans, the relative distribution was calculated. The rhEpo-Fc concentration before papain digestion was 2.2 nmol. The N-glycans were identified by comparison to oligosaccharide standards. The names of the different glycans are listed in Fig. 5. The maximum SD of three independent analyses of the N-glycans amounted ±0.92%.

molecule was also published [18,19]. Concerning the volumetric titers obtained with different hosts no data were available, thus only values published for rhEpo could be taken into account. For rhEpo producing CHO cells volumetric titer in the range of 200–3000 pmol/ml were found, whereas higher titers were only seen in optimized large scale production systems [20–22]. However, our titers were absolutely comparable with titers published for the rhEpo in small scale cultivation, which was promising for a large scale production.

Isoelectric focusing resulted in 15 isoforms in the range of pH 4.6–6.5. The most intensive band representing the predominant isoform focused at about pH 5.2 (Fig. 4). RhEpo-Fc fusion protein expressed in BHK cells, showed an isoform pattern in the range of pH 4.6–5.0 with fainter bands between pH 5.0 and 6.0 whereas, no further quantitative data were available. In contrast, rhEpo-Fc, expressed in NS/0 cells, migrated with an isoelectric pattern between pH 5.3 and 6.5, the most intense band focused at about pH 6.0–6.1 [12]. Thus, expression of rhEpo-Fc in BHK cells resulted in a significantly more acidic product than identical or similar products expressed in NS/0 cells. For this reason, the more basic isoform distribution of rhEpo-Fc, expressed in CHO cells, correlates with the average sialylation degree measured in our studies and in accordance to the sialylation degree obtained for erythropoietin expressed in CHO cells.

In theory, one rhEpo-Fc molecule can reach a total of 28 sialic acids/molecule. The extent of sialylation is variable from site to site, protein to protein and can be dependent on cell culture conditions, cell types and in particular cell clones that are used. The N-acetylneuraminic acid concentration we found was between 10 and 14 mol/mol rhEpo-Fc. This means that 35–50% of all potential sialylation sides are sialylated. Higher sialylation degrees have been reported by Gillies et al. [12]. A population of highly sialylated rhEpo-Fc proteins, expressed in BHK cells, had an average of 11–28 N-acetylneuraminic acid residues per purified rhEpo-Fc fusion protein. In accordance to these data, it was reported that the N-acetylneuraminic acid concentration of the recombinant Epo produced in BHK-21 cells was also increased compared to CHO cells [23]. In contrast, a population of rhEpo-Fc proteins expressed in NS/0 cells appeared to had an average N-acetylneuraminic acid amount of 10 per molecule [12]. Therefore, our Protein A purified rhEpo-Fc sample, which was expressed in CHO cells, represents a medial sialylation degree.

The mixture of neutral and charged N-glycans released from rhEpo-Fc was separated into approximately 16 N-glycan structures (Fig. 6). These were identified by normal-phase HPLC on an Amide-80 column. The polar functional groups of the column interact with the hydroxyl groups on the oligosaccharide, thus neutral and charged oligosaccharides can be separated simultaneously on the basis of hydrophilic interaction. In principle, large oligosaccharides were eluted by increasing concentration of ammonium acetate. For this reason, neutral oligosaccharides were eluted first, followed by charged oligosaccharides. The N-glycans were identified by the retention times of defined oligosaccharide standards and afterwards verified by sequential glycan cleavage (Fig. 7). Additionally, relative amounts of N-glycans were quantified by measuring the peak areas and expressed as percentage of the total released N-glycans. This calculation assumes that each N-glycan molecule is labeled with one molecule of 4-ABA and that each molecule of glycan-bound 4-ABA contributes the same amount of fluorescence [24].

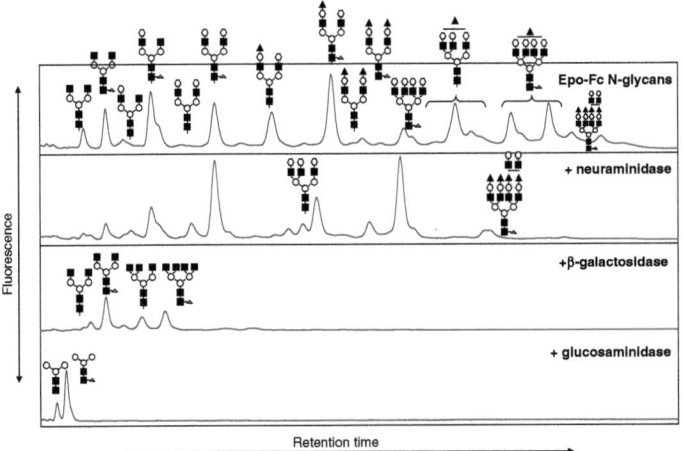

Fig. 8. Sequential exoglycosidase digestions of 4-ABA labeled N-glycans released from rhEpo-Fc separated on normal-phase HPLC. The N-glycan concentration was approximately 440 pmol before cleavage. The chromatograms show the results of the sequential digestion, where the complexity of the N-glycan pattern is successively reduced.

As expected, the fusion protein showed a very heterogeneous N-glycan pattern. The four main detected oligosaccharide structures ranged between 13 and 18%, whereas A4 was the dominating structure with the highest amount (18%), followed by A3 as well as A1F (15%) and NA1F (13%). Additionally, rhEpo-Fc showed a highly fucosylated N-glycan pattern, about 60% of all complex N-glycans were fucosylated. The comparison with the rhEpo-Fc glycan pattern observed by other researchers is not possible, because neither in the literature nor in available patents detailed data were found.

However, this kind of analysis gave no information about the glycosylation of the subunits, thus detailed information only could be obtained after protein cleavage. Because the gene sequence of the fusion protein lacks a specific cleavage site, the rhEpo subunit was separated from the rhFc subunit by papain digestion. According to Adamczyk [25], papain should preferentially cleave in the hinge region of the rhFc subunit. The presence of reducing agents, usually cysteine, was necessary for enzyme activation with the disadvantage of undesirable cleavage of labile disulfide bridges. The cleavage at the N-terminal side of the hinge region resulted in double chain rhFc fragments, while cleavage at the C-terminal side of the hinge region would produce rhFc monomer fragments [25]. This fact may play a critical role for further structural analysis, but in our approach cleavage did not alter the analysis of N-glycan structures. Additionally, separation of the rhEpo and rhFc subunit by Protein A affinity chromatography were not affected since Protein A interacted with His435 of each rhFc chain through hydrophobic contacts [26,27].

We could clearly show that the rhEpo subunit as well as the rhFc subunit of the recombinant rhEpo-Fc molecule was glycosylated (Fig. 7). The predominant detected N-glycan structures of the rhFc subunit were NA1F (24%) and NA1 (20%) followed by NGA2F, NA2 and NA2F (each 12%). Furthermore, 87% of all rhFc N-glycans were neutral. All other N-glycan structures were detected in small amounts, each <10%. These findings are consistent with the results of other studies for the characterization of antibody glycosylation, which have shown that the majority of the N-glycans were NA2F, NA1F and NGA2F [28,29]. N-Glycans of human IgG$_1$ subclasses also contain bisecting branches. However, CHO cells are not able to express bisecting N-acetylglucosamine transferase, which could be confirmed by our data [30].

The predominate detected N-glycans of the rhEpo subunit were completely different to the rhFc subunit. A4 (34%), A3 (20%) and A1F (14%) were the dominating glycans of the rhEpo subunit. Approximately 11% of all rhEpo N-glycans were neutral. All other N-glycan structures were detected in small amounts, each <10%. These findings are also consistent with the results of other studies, which have shown that the majority of the N-glycans of rhEpo were tetraantennary, but they also show that not all of them were fully sialylated [31,32]. Yang and Butler studied the N-glycan structures of affinity purified recombinant Epo by anion exchange chromatography [33]. Their results show that the major oligosaccharide structures were tetrasialylated (46%) followed by the trisialylated (28%), disialylated (14%) and monosialylated (12%) structures. Yuen et al. studied the

N-glycan structure of different ion exchange purified recombinant Epo samples by normal-phase HPLC [24]. With the exception of a cation exchange purified sample, which contained the more basic isoforms, all other samples contained between 31 and 57% of the tetrasialylated tetraantennary oligosaccharides. These results are comparable with our findings.

For sequential digestion procedure the total N-glycan pool of rhEpo-Fc was treated with different exoglycosidases (neuraminidase from *A. ureafaciens*, β-galactosidase and glucosaminidase from *S. pneumoniae*) (Fig. 8). This analysis was performed to confirm the identity of standards and the quality of the measured data.

Conclusion

In this project, we could demonstrate that rhEpo-Fc fusion protein can be overexpressed in CHO cells in amounts comparable to the glycoprotein rhEpo itself. Purification was performed with affinity chromatography to get a representative capture pool. The same technique was used to prepare fractions containing either the rhEpo or the rhFc subunit, which enables the characterization of both subunits. All used methods for the biochemical characterization were optimized to obtain more quantitative than qualitative information.

The current study showed that no significant differences in the core glycan structures of papain digested rhEpo-Fc compared to crude rhEpo and human IgG glycans were found. The direct comparison with other rhEpo-Fc fusion proteins failed, because no appropriate data were found in the literature. However, evident differences were detectable in the sialylation degree compared to other production systems, such as NS/0 and BHK.

Acknowledgments

The authors thank Annalisa Lasagna for their analytical support. This research was kindly funded by ACBT (Austrian Center of Biopharmaceutical Technology), a competence center supported by the Federal Ministry of Economy and Labour and the federal states of Vienna and Tyrol.

References

[1] Y. Sudo, M. Yamamoto, A. Orikasa, H. Inomata, Y. Takeshita, Fused Protein, EP1088888, Germany, 1999.
[2] A.J. Bitonti, J.A. Dumont, S.C. Low, R.T. Peters, K.E. Kropp, V.J. Palombella, J.M. Stattel, Y. Lu, C.A. Tan, J.J. Song, A.M. Garcia, N.E. Simister, G.M. Spiekermann, W.I. Lencer, R.S. Blumberg, Pulmonary delivery of an erythropoietin Fc fusion protein in non-human primates through an immunoglobulin transport pathway, Proc. Natl. Acad. Sci. USA 101 (2004) 9763–9768.
[3] S.C. Low, S.L. Nunes, A.J. Bitonti, J.A. Dumont, Oral and pulmonary delivery of FSH-Fc fusion proteins via neonatal Fc receptor-mediated transcytosis, Hum. Reprod. 20 (2005) 1805–1813.
[4] T. Miyake, C.K. Kung, E. Goldwasser, Purification of human erythropoietin, J. Biol. Chem. 252 (1977) 5558–5564.
[5] J.M. Davis, T. Arakawa, T.W. Strickland, D.A. Yphantis, Characterization of recombinant human erythropoietin produced in Chinese hamster ovary cells, Biochemistry 26 (1987) 2633–2638.
[6] M.A. Recny, H.A. Scoble, Y. Kim, Structural characterization of natural human urinary and recombinant DNA-derived erythropoietin. Identification of des-arginine 166 erythropoietin, J. Biol. Chem. 262 (1987) 17156–17163.
[7] P.H. Lai, R. Everett, F.F. Wang, T. Arakawa, E. Goldwasser, Structural characterization of human erythropoietin, J. Biol. Chem. 261 (1986) 3116–3121.
[8] J.K. Browne, A.M. Cohen, J.C. Egrie, P.H. Lai, F.K. Lin, T. Strickland, E. Watson, N. Stebbing, Erythropoietin: gene cloning, protein structure, and biological properties, Cold Spring Harb. Symp. Quant. Biol. 51 (1986) 693–702.
[9] J.C. Egrie, T.W. Strickland, J. Lane, K. Aoki, A.M. Cohen, R. Smalling, G. Trail, F.K. Lin, J.K. Browne, D.K. Hines, Characterization and biological effects of recombinant human erythropoietin, Immunobiology 172 (1986) 213–224.
[10] R.S. Rush, P.L. Derby, T.W. Strickland, M.F. Rohde, Peptide mapping and evaluation of glycopeptide microheterogeneity derived from endoproteinase digestion of erythropoietin by affinity high-performance capillary electrophoresis, Anal. Chem. 65 (1993) 1834–1842.
[11] R.S. Rush, P.L. Derby, D.M. Smith, C. Merry, G. Rogers, M.F. Rohde, V. Katta, Microheterogeneity of erythropoietin carbohydrate structure, Anal. Chem. 67 (1995) 1442–1452.
[12] S.D. Gillies, J. Way, K.M. Lo, FC-Erythropoietin fusion protein with improved pharmacokinetics, US2005/0202538 A1, USA, 2005.
[13] W. Diezel, G. Kopperschlager, E. Hofmann, An improved procedure for protein staining in polyacrylamide gels with a new type of Coomassie brilliant blue, Anal. Biochem. 48 (1972) 617–620.
[14] P.G. Righetti, Immobilized pH Gradients: Theory and Methodology, Elsevier Biomedical Press, Amsterdam, 1990.
[15] R. Westermeier, Electrophoresis in Practice, Wiley-VCH, Weinheim, 2005.
[16] D.J. Capon, S.M. Chamow, J. Mordenti, S.A. Marsters, T. Gregory, H. Mitsuya, R.A. Byrn, C. Lucas, F.M. Wurm, J.E. Groopman, S. Broder, D.H. Smith, Designing CD4 immunoadhesins for AIDS therapy, Nature 337 (1989) 525–531.
[17] J.C. Way, S. Lauder, B. Brunkhorst, S.-M. Kong, A. Qi, G. Webster, I. Campbell, S. McKenzie, Y. Lan, B. Marelli, L.A. Nguyen, S. Degon, K.-M. Lo, S.D. Gillies, Improvement of Fc-erythropoietin structure and pharmacokinetics by modification at a disulfide bond, Protein Eng. Des. Sel. 18 (2005) 111–118.
[18] A.J. Sytkowski, E.D. Lunn, K.L. Davis, L. Feldman, S. Siekman, Human erythropoietin dimers with markedly enhanced in vivo activity, PNAS 95 (1998) 1184–1188.
[19] A.M. DePaolis, J.V. Advani, B.G. Sharma, Characterization of erythropoietin dimerization, J. Pharm. Sci. 84 (1995) 1280–1284.
[20] N.Y. Kim, J.H. Kim, H.J. Kim, Effect of low adapted temperature and medium composition on growth and erythropoietin (EPO) production by Chinese hamster ovary cells, Arch. Pharm. Res. 28 (2005) 220–226.
[21] F. LeFloch, B. Tessier, S. Chenuet, J.-M. Guillaume, P. Cans, A. Marc, J.-L. Goergen, HPCE monitoring of the N-glycosylation pattern and sialylation of murine erythropoietin produced by CHO cells in batch processes, Biotechnol. Prog. 20 (2004) 864–871.
[22] G.M. Lee, E.J. Kim, N.S. Kim, S.K. Yoon, Y.H. Ahn, J.Y. Song, Development of a serum-free medium for the production of erythropoietin by suspension culture of recombinant Chinese hamster ovary cells using a statistical design, J. Biotechnol. 69 (1999) 85–93.
[23] M. Nimtz, W. Martin, V. Wray, K.D. Kloppel, J. Augustin, H.S. Conradt, Structures of sialylated oligosaccharides of human erythropoietin expressed in recombinant BHK-21 cells, Eur. J. Biochem. 213 (1993) 39–56.

[24] C.T. Yuen, C.K. Gee, C. Jones, High-performance liquid chromatographic profiling of fluorescent labelled N-glycans on glycoproteins, Biomed. Chromatogr. 16 (2002) 247–254.
[25] M. Adamczyk, J.C. Gebler, J. Wu, Papain digestion of different mouse IgG subclasses as studied by electrospray mass spectrometry, J. Immunol. Methods 237 (2000) 95–104.
[26] A.E. Sauer-Eriksson, G.J. Kleywegt, M. Uhlen, T.A. Jones, Crystal structure of the C2 fragment of streptococcal protein G in complex with the Fc domain of human IgG, Structure 3 (1995) 265–278.
[27] L. Jendeberg, P. Nilsson, A. Larsson, P. Denker, M. Uhlen, B. Nilsson, P.A. Nygren, Engineering of Fc(1) and Fc(3) from human immunoglobulin G to analyse subclass specificity for staphylococcal protein A, J. Immunol. Methods 201 (1997) 25–34.
[28] R. Jefferis, J. Lund, H. Mizutani, H. Nakagawa, Y. Kawazoe, Y. Arata, N. Takahashi, A comparative study of the N-linked oligosaccharide structures of human IgG subclass proteins, Biochem. J. 268 (1990) 529–537.
[29] G.R. Guile, P.M. Rudd, D.R. Wing, S.B. Prime, R.A. Dwek, A rapid high-resolution high-performance liquid chromatographic method for separating glycan mixtures and analyzing oligosaccharide profiles, Anal. Biochem. 240 (1996) 210–226.
[30] J.P. Kamerling, Carbohydrate features of recombinant human glycoproteins, Biotechnol. Appl. 13 (1996) 167–180.
[31] K. Kanazawa, K. Ashida, M. Itoh, H. Nagai, H. Sasaki, M. Fukuda, Establishment of a method for mapping of N-linked oligosaccharides and its use to analyze industrially produced recombinant erythropoietin, Biol. Pharm. Bull. 22 (1999) 339–346.
[32] C.H. Hokke, A.A. Bergwerff, G.W. Van Dedem, J.P. Kamerling, J.F. Vliegenthart, Structural analysis of the sialylated N- and O-linked carbohydrate chains of recombinant human erythropoietin expressed in Chinese hamster ovary cells. Sialylation patterns and branch location of dimeric N-acetyllactosamine units, Eur. J. Biochem. 228 (1995) 981–1008.
[33] M. Yang, M. Butler, Effects of ammonia and glucosamine on the heterogeneity of erythropoietin glycoforms, Biotechnol. Prog. 18 (2002) 129–138.

Journal of Pharmaceutical and Biomedical Analysis 42 (2006) 322–327

www.elsevier.com/locate/jpba

A novel strategy for quantitative isoform detection directly performed from culture supernatant

Kornelia Schriebl [a,*], Evelyn Trummer [a,b], Robert Weik [c], Dethardt Müller [b], Renate Kunert [b], Christine Lattenmayer [a,b], Hermann Katinger [b,c], Karola Vorauer-Uhl [b]

[a] *Austrian Center of Biopharmaceutical Technology, Muthgasse 18, A-1190 Vienna, Austria*
[b] *Institute of Applied Microbiology, Department of Biotechnology, University of Natural Resources and Applied Life Science, Muthgasse 18, A-1190 Vienna, Austria*
[c] *Polymun Scientific GmbH, Nussdorferlaende 11, A-1190 Vienna, Austria*

Received 27 January 2006; received in revised form 10 April 2006; accepted 11 April 2006
Available online 9 June 2006

Abstract

Currently, one of the most used techniques for the determination of isoform pattern analysis is isoelectric focusing. Routinely, this is performed by immunoblotting. Blotting of proteins after isoelectric focusing on IPG gels may cause several problems, such as protein loss by the blotting itself and band broadening, in some cases the immunostaining with antibodies might be problematic. In the present study, an alternative isoform prestaining method with CyDye fluors is presented. For this approach, a highly glycosylated fusion protein, Epo-Fc, was used consisting of two recombinant human erythropoietin attached to the Fc part of a human IgG$_1$ molecule. By using CyDye fluors, up to three samples can be focused on the same lane under identical electrophoretic conditions. A fundamental benefit of this technique is the ability to perform quantitative isoform pattern analysis directly from serum-free culture supernatant.
© 2006 Elsevier B.V. All rights reserved.

Keywords: Fusion protein Epo-Fc; CHO cells; Isoelectric focusing; Isoform pattern; CyDye fluors

1. Introduction

Routinely, visualization of isoform pattern from culture supernatant and purified protein were performed on IPG gels by immunoblotting [1,2]. However, some drawbacks are inherent to these techniques. Besides the technical deficiencies, such as protein loss by the blotting itself and band broadening, in some cases the immunostaining with antibodies might be problematic. Even though specific antibodies are available, greater or lesser extent of background staining and insufficient linearity prevent quantification. Due to these drawbacks, immunoblotting is a time-consuming qualitative method. However, with the development of new imaging systems, protein detection and quantitation methods based on fluorescent staining and/or labeling provide a promising alternative. Depending on the chemical and phys-

ical nature of the fluorophore, they have a comparatively wide linear dynamic range (>10^3) [3]. CyDye fluors, mainly used for 2D difference gel electrophoresis (2D DIGE), allow more accurate and sensitive quantitative protein studies. Currently, three different CyDye fluors (Cy 5, Cy 3 and Cy 2) are available from GE Healthcare (Sweden). The fluorophores are structurally similar and undergo nucleophilic substitutions with the ε-amino group of lysine residues forming an amide. The fluorescent dyes have very similar molecular masses and are positively charged to match the charge that is replaced on the lysine residue. The charge and mass matching ensure that all samples essentially co-migrate to the same point during electrophoresis. In the labeling reaction, the dye:protein ratio is low. This ensures that protein molecules are only labeled with a single dye molecule [4]. In fact, only 1–2% of lysine residues in the proteins are fluorescently modified, so that the solubility of the labeled proteins is maintained during electrophoresis [5]. Due to the availability of three different CyDye fluors with similar attributes, up to three different samples can be analyzed on the same lane. Thus, by

* Corresponding author. Tel.: +43 01 36006 6229; fax: +43 01 3697615.
E-mail address: kschriebl@gmx.at (K. Schriebl).

0731-7085/$ – see front matter © 2006 Elsevier B.V. All rights reserved.
doi:10.1016/j.jpba.2006.04.021

this technique spiking experiments with reference standards or other samples can be performed simultaneously.

To evaluate the suitability of this prestaining technique for one-dimensional IEF, all experiments were performed with our fusion protein Epo-Fc, which consists of two recombinant human erythropoietin (rHuEpo) attached to the Fc part of a human IgG$_1$ molecule. Epo-Fc comprises of 798 amino acids and has a molecular weight of about 112 kDa, of which about 89 kDa is contributed by the polypeptide chain itself. For the characterization of posttranslational modifications, respectively, the analysis of isoforms, the sialylated glycan structures presented by the Epo molecules are primarily responsible. HuEpo is a 30.4 kDa glycoprotein hormone produced by the kidney in adult humans [6,7]. It consists of a 165 amino acid single polypeptide chain containing two disulfide bonds [8,9] and has three N-linked (Asn-24, Asn-38, Asn-83) and one O-linked (Ser-126) sugar chain [10,11]. Its microheterogeneity is related to the charged carbohydrate moiety of the protein and is studied extensively [12,13]. The microheterogeneity of HuEpo is seen on the N-linked carbohydrate chains, where the oligosaccharide may contain bi-, tri- and tetra-antennae, each of which is typically terminated with the negatively charged sialic acid molecule. With the exception of sialic acid, all sugar moieties are neutral. As a consequence of the variability of sugar structures, the number of sialic acids molecules varies and give rise to HuEpo isoforms with differences in charge. Therefore, the electrophoretic mobility and the isoelectric point (pI) or pattern are strongly influenced by the degree of sialylation of polysaccharide chains. Due to this, microheterogeneity not only determines the biochemical properties of this molecule, but also its biological activity, respectively, half life time. For instance, the amount of terminal sialylated tetraantennary structures prevents rapid clearance of HuEpo by the liver in vivo [14,15]. Furthermore, the molecule itself is stabilized by the carbohydrate structure, whereas the non-glycosylated form is less stable than all other asialic or sialic variants [16]. Therefore, analysis of the isoform pattern is an indispensable part in quality control. As described above, our fusion protein consists of two rHuEpo attached to the Fc part of a human IgG$_1$ molecule. Thus, the microheterogeneity of the fusion protein Epo-Fc is manifested in a complex isoform pattern.

The aim of this study was to evaluate the potency of this method using our complex model protein Epo-Fc. Optimization for optimal protein amount, reproducibility and sample preparation was performed. Thereafter, the method was evaluated for serum-free culture supernatants. Therefore, evaluation was accomplished with three different purified protein samples and their corresponding culture supernatant followed by simultaneous analysis of an in-house standard and selected samples.

2. Materials and methods

2.1. Materials

All reagents used were electrophoresis grade. CyDye DIGE Fluors (minimal dyes) for Ettan DIGE, dimethylformamide (DMF), 2D Clean-Up Kit, Pharmalyte (pH 3–10), dithiothreitol (DTT), GelBond-PAG film and Protein A Sepharose FF were purchased from GE Healthcare (Sweden). Proto Gel Ultrapure was purchased from National Diagnostics (USA). Tetramethylethylenediamine (TEMED), acrylamido buffer solution "Immobiline" pK 3.6, 4.6, 6.2, 7.0, 8.5 and 9.3, lysine, Tris, glycine, bromophenole blue, ammoniumperoxodisulfate (APS) and 3-[(3-cholamidopropyl)dimethylammonio] propanesulfonic acid (CHAPS) were from Fluka (Switzerland). Glycerol and urea were from Merck (Germany).

2.2. Cell line, culture medium and cultivation

Dihydrofolate reductase-deficient Chinese Hamster Ovary cells (DUKX-B11, ATCC CRL-9096) were co-transfected with genes for fusion protein Epo-Fc and dihydrofolate reductase (DHFR). Transfected cells were selected for growth in the presence of 0.096 μM MTX. After revitalization from a research cell bank, cells were cultivated in suspension in Dulbecco's Modified Eagle's Medium DMEM/HAM'S F-12 (1:1 mixture), supplemented with 0.58 g/l L-glutamine, an in-house developed protein-free supplement (proprietary formulation), 0.25% soy peptone, 0.1% Pluronic F68 and 0.096 μM MTX. Within 2 weeks the CHO cells were adapted in spinner flasks (TECHNE, UK) to an in-house developed production medium (Polymun Scientific GmbH, Austria) of proprietary formulation. Repeated batch cultivation at different pH and temperature levels were performed in a Sixfors multireactor system (Infors, Switzerland).

2.3. Purification of Epo-Fc

The secreted Epo-Fc from CHO cells was purified by immunoaffinity on Protein A Sepharose FF (GE Healthcare, Sweden). Cell-free culture supernatant samples containing Epo-Fc were adjusted to pH 8.5 with Tris (pH 9.0; 1 M) and loaded onto a 314 μl HR 5 column at a flow-rate of 0.2 ml/min (60 cm/h). After extensive washing of the column with Tris (0.025 M)–NaCl (0.15 M) (pH* 8.5), Epo-Fc was eluted with glycine–HCl (pH 3.5; 0.1 M) with a flow-rate adjusted to 0.1 ml/min (30 cm/h). The eluted fraction was rapidly neutralized with Tris (pH 9.0; 1 M). After elution, the column was regenerated with glycine–HCl (pH 2.5; 0.1 M).

2.4. Isoelectric focusing of Epo-Fc

2.4.1. Sample preparation and labeling

During IEF the salt concentration should always be kept below 0.05 M. For that purpose, purified and non-purified Epo-Fc samples were treated with 2D Clean Up Kit according to the kit instructions and resuspended in Tris (0.03 M), urea (3 M), CHAPS (4% (w/v)). Thereafter, 4 pmol/8 μl of Epo-Fc sample was minimally labeled with 1 μl of Cy 5 and Cy 3 (3.5 pmol/μl) on ice for 30 min in the dark. The labeling reaction was quenched by the addition of 1 μl lysine (0.01 M) and subsequently incubated on ice for 10 min in the dark. Following the labeling reaction, the Cy 5 labeled sample was mixed with the Cy 3 labeled sample and adjusted with CHAPS (4% (w/v)) to the appropriate volume (40 μl), or 20 μl if just one sample (Cy 5 or Cy 3) was focused.

2.4.2. Gel preparation

Preparation of the linear IPG gel (pH 4–8) was performed as previously described [17,18]. The IPG gel was rehydrated for 2 h in a solution containing urea (6 M), DTT (0.01 M), CHAPS (2% (w/w)), bromophenole blue 0.1% (0.4% (w/w)) and Pharmalyte 3–10 (2% (w/w)). After rehydration, the linear IPG gel was scratched into 2 cm wide strips to prevent sample interfering during focusing.

2.4.3. Isoelectric focusing

IEF of labeled samples was performed on a Multiphor II flat bed electrophoresis unit (GE Healthcare, Sweden) [17] with the following settings: 15 °C, 2.5 mA, 5 W, 0.075 kVh (0.75 h, 100 V, step-n-hold), 0.375 kVh (0.75 h, 500 V, step-n-hold), 0.500 kVh (0.5 h, 1000 V, step-n-hold) and 72 kVh (18 h, 4000 V, step-n-hold). After focusing, the gel was washed two times with water for 5 min each in order to remove paraffin. Thereafter, the gel was immediately scanned using the Typhoon 9400™ variable mode imager (GE Healthcare, Sweden). Gels were placed gel side-down on the imager for optimal detection.

2.4.4. Image acquisition

The Cy 5 images were scanned using a 633 nm laser and an emission filter of 670 nm bandpass (BP) 30. Cy 3 images were scanned using a 532 nm laser and an emission filter of 580 nm BP 30. The photomultiplier tube (PMT) voltage was adjusted for each channel (Cy 5 and Cy 3) in preliminary low-resolution scans to give maximum pixel values but keep the signals below the saturation level. These settings were subsequently used for high resolution (100 μm) scans for all gels. Image analysis was carried out with the Image Quant TL v2003 software (GE Healthcare, Sweden).

3. Results

The microheterogeneity of Epo-Fc is mainly dependent on the presence of terminal sialic acid residues on glycan structures. To determine the complex isoform pattern of Epo-Fc, IEF was performed with prelabeled samples on IPG gels (pH 4–8).

3.1. Reproducibility

Due to the fact that a certified reference standard was not available for our fusion protein we used a representative purified Epo-Fc sample as internal reference standard (IRS). To evaluate the suitability, respectively, reproducibility of the prestaining method, we were focusing the IRS on four independent lanes under the same electrophoretic conditions. Therefore, the IRS was treated with 2D Clean Up Kit, labeled with Cy 3 and applied onto the gel. After scanning, isoform interpretation was performed by Image Quant TL v2003 software (GE Healthcare, Sweden). The mean values of the percentage distribution of the

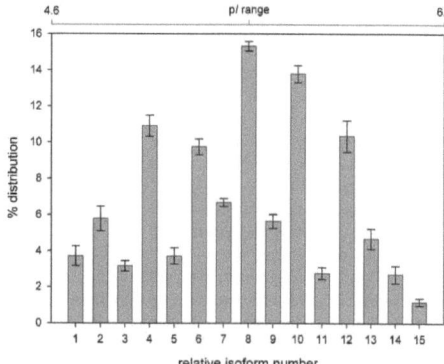

Fig. 1. Reproducibility of the isoform pattern of the IRS. Epo-Fc concentration was 4 pmol. The maximum S.D. of four independent analyses of several corresponding isoforms amounted ±0.9%.

IRS isoforms are shown in Fig. 1. Based on these results, we analyzed different Epo-Fc samples.

3.2. Isoform pattern of non-purified and corresponding purified Epo-Fc samples

To evaluate the suitability for the analysis of cell-free culture supernatants we analyzed various serum-free culture supernatants compared to corresponding purified samples. Cells were cultivated under different temperature and pH values. All samples were treated with 2D Clean Up Kit, labeled with Cy 5 and applied onto the gel. After scanning, isoform interpretation was performed by Image Quant TL v2003 (GE Healthcare, Sweden).

Cells cultivated at 30 °C, pH 6.9 in a Sixfors multireactor system were prepared as described above. The percentage distribution of the isoforms obtained with serum-free culture supernatant and the corresponding purified protein are shown in Fig. 2A. The isoform pattern of both samples resulted in similar distributions. Samples were analyzed in triplicates and the standard deviation of each isoform was calculated. As shown in Fig. 2A–C, the standard deviations are randomly distributed and not dependent on the magnitude of the measurements. The maximum difference between the corresponding isoforms, purified versus non-purified, was less than ±0.8%.

To verify the methodology itself, analyses were performed as described above with two other samples. Thereby, serum-free culture supernatants cultivated under different conditions were used.

In Fig. 2B and C the percentage distributions of the isoforms of the serum-free culture supernatant and the corresponding purified protein of two other samples, which were cultivated at 30 °C, pH 7.1 and 37 °C, pH 6.9 in a Sixfors multireactor system, are shown. Samples were analyzed in triplicates and the standard deviation of each isoform was calculated. The isoform pattern of both corresponding samples resulted in similar distributions. The maximum difference between the corresponding

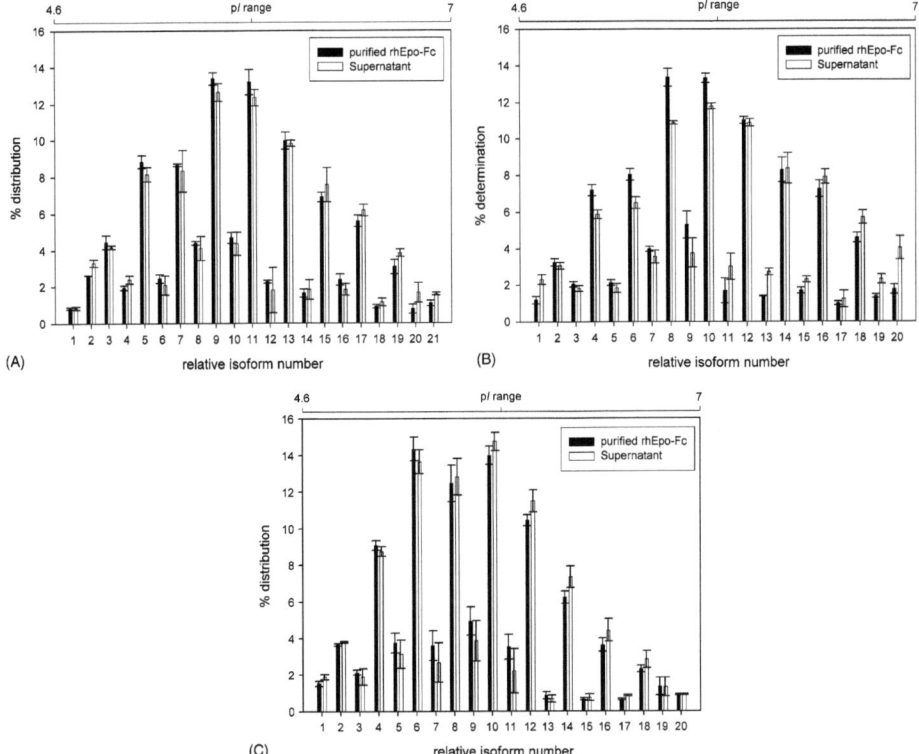

Fig. 2. Isoform pattern of a non-purified and the corresponding purified Epo-Fc sample under different culture conditions (A) 30 °C, pH 6.9, (B) 30 °C, pH 7.1, (C) 37 °C, pH 6.9, directly performed in the serum-free culture supernatant compared to the purified protein. The Epo-Fc concentration was 4 pmol. S.D. is shown with error bars of each isoform, calculated from triplicate analysis.

isoforms, purified versus non-purified, was less than ±2.4% for the sample cultivated at 30 °C, pH 7.1 and less than ±1.4% for the sample cultivated at 37 °C, pH 6.9.

Additionally, IRS and serum-free culture supernatant were prestained with different CyDye fluors. The IRS was labeled with Cy 3 and the serum-free culture supernatant was labeled with Cy 5. After labeling, both samples were mixed, applied onto the gel and separated on the same lane under identical electrophoretic conditions. After focusing, each sample was scanned at the appropriate wavelengths of the respective CyDye fluor. The isoform patterns of IRS, serum-free culture supernatant and the overlay are shown in Fig. 3. Both samples showed similar isoforms, which is represented by the yellow colored bands of the overlay. Additional, green or red bands in the overlay would demonstrate that the samples consisted of varying isoforms. In our case, the IRS and the sample showed similar isoforms except of the basic isoforms of the sample, which is shown by the appearance of red bands at the basic side of the gel. By this matching technique, each sample can be quantitatively compared to the selected IRS.

4. Discussion

Isoelectric focusing, to estimate biochemical properties of proteins, determined by the amino acid sequence and additional charges of posttranslational modifications, is a state of the art technology for many therapeutically used recombinant proteins. In parallel, the electrophoresis technology with high resolution gels and more efficient staining methods were established. Particularly, the establishment of 2D electrophoresis for entire protein separation necessitates more sensitive staining. For this purpose, fluorescent dyes with appropriate quality, such as the CyDyes, were developed. Due to their characteristics, CyDye fluors afford a sensitive prestaining of proteins. By the minimal labeling, the mobility of proteins and peptides during IEF is not affected and broad linear range quantification is feasi-

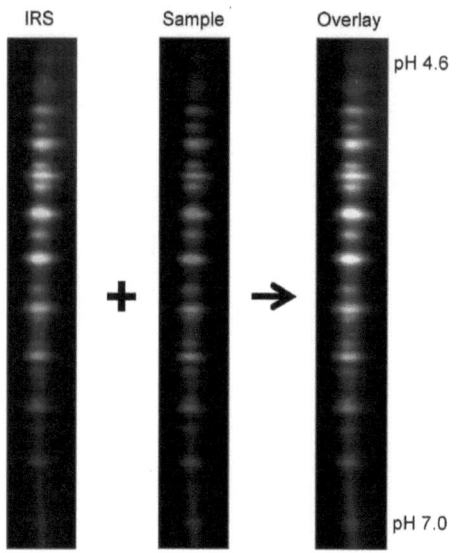

Fig. 3. Imaging. Isoform pattern of two Epo-Fc samples, directly performed from the serum-free culture supernatant by CyDye fluors. The IRS was labeled with Cy 3 and the serum-free culture supernatant was labeled with Cy 5, mixed and separated on the same lane under identical electrophoretic conditions. The isoform pattern at the right represented the overlay of the IRS and the serum-free culture supernatant. The Epo-Fc concentration of each sample was 4 pmol.

ble. High sensitivity (low detection limit), high linear dynamic range (excellent quantitative accuracy) and reproducibility feature them as eligible candidates for all electrophoretic methods [3].

In our approach, we evaluated the use of Cy 5 and Cy 3 CyDyes regarding reproducibility, suitability for crude cell-free culture supernatants and overlaying method. Therefore, we started to prove the reproducibility and suitability of the prestaining method exemplified on the complex protein Epo-Fc. Our results, obtained with four independent analyses, showed a maximum S.D. of ±0.9% corresponding isoforms. Based on these results, we analyzed different Epo-Fc samples. In this regard, purified as well as protein in culture supernatants were analyzed. For this approach, different culture supernatants containing Epo-Fc were harvested and a part of it was purified by affinity chromatography. All different culture supernatants were generated in serum-free media but under different culture conditions. The isoform pattern of the serum-free culture supernatant and the corresponding purified protein of all three analyzed samples showed comparable results. The maximum differences between the corresponding isoforms of all three samples ranged between ±0.8 and ±2.4%. Based on these results, we could demonstrate that this technique, including 2D Clean Up Kit and labeling with CyDye fluors, is suitable for protein characterization directly from culture supernatants. However,

supernatants contain greater or lesser extent of various host proteins. Therefore, characterization of the protein of interest is only possible when these proteins are presented in small amounts and/or the protein of interest dominates. Nevertheless, CyDye fluors are very sensitive, detection limits of about 0.25–0.95 ng are published for 1D SDS-PAGE using protein standards [4]. Different to 1D SDS-PAGE, heterogeneous proteins, separated on IPG gels, do not migrate to one single spot, but were separated in various isoforms with individual amounts. In the case of Epo-Fc, 15–21 isoforms are detectable on IPG gels with a relative distribution in the range of about 0.5–14%, thus the total protein amount loaded onto the gel must be adjusted accordingly.

By the use of non-purified protein solutions in combination with fluorescence techniques faster analysis of protein isoforms in cell screening and fermentation is possible, avoiding prior time-consuming protein purification.

Furthermore, we directly compared the serum-free culture supernatant and the IRS by labeling them with Cy 5 and Cy 3 which were separated on the same lane and scanned at the appropriate wavelengths. With this technique up to three samples can be focused on the same lane under identical electrophoretic conditions. A fundamental benefit of this overlaying technique is the ability to co-detect and compare each sample in-gel with an internal standard. This internal standard can be used for normalization of the isoforms across all gels. With this approach, the experimental variation is further reduced and the accuracy of quantification is increased.

5. Conclusion

In summary, IEF in combination with 2D Clean Up Kit and CyDye fluors can be directly performed with protein/serum-free culture supernatants as well as purified proteins. Therefore, this technology provides a useful tool for clone screening and product quality analysis. Apart from our studies, it is advisable for optimization to analyze culture supernatants in comparison with the purified protein of interest to prove matrix influences. In the present study, 4 pmol Epo-Fc was loaded to quantify the relative isoform distribution, whereas the individual isoforms were in the range of 0.5–14%. Depending on the characteristics of the protein of interest, for instance the number of isoforms, the protein amount must be adjusted accordingly. Therefore, the protein concentration, corresponding with the detection limit, respectively, quantification limit, need to be optimized. However, it should be emphasized that depending on the linear dynamic range of a dye, there is always a compromise between the detection of minor and saturated isoforms.

Acknowledgements

The authors would like to thank Annalisa Lasagna for technical support with IEF. This research was kindly funded by ACBT (Austrian Center of Biopharmaceutical Technology), a competence center supported by the Federal Ministry of Economy and Labour and the federal states of Vienna and Tyrol.

References

[1] F. Lasne, J. De Ceaurriz, Nature 405 (2000) 635.
[2] F. Lasne, L. Martin, N. Crepin, J. de Ceaurriz, Anal. Biochem. 311 (2002) 119–126.
[3] A. Gorg, W. Weiss, M.J. Dunn, Proteomics 4 (2004) 3665–3685.
[4] R. Tonge, J. Shaw, B. Middleton, R. Rowlinson, S. Rayner, J. Young, F. Pognan, E. Hawkins, I. Currie, M. Davison, Proteomics 1 (2001) 377–396.
[5] M. Unlu, M.E. Morgan, J.S. Minden, Electrophoresis 18 (1997) 2071–2077.
[6] T. Miyake, C.K. Kung, E. Goldwasser, J. Biol. Chem. 252 (1977) 5558–5564.
[7] J.M. Davis, T. Arakawa, T.W. Strickland, D.A. Yphantis, Biochemistry 26 (1987) 2633–2638.
[8] M.A. Recny, H.A. Scoble, Y. Kim, J. Biol. Chem. 262 (1987) 17156–17163.
[9] P.H. Lai, R. Everett, F.F. Wang, T. Arakawa, E. Goldwasser, J. Biol. Chem. 261 (1986) 3116–3121.
[10] J.K. Browne, A.M. Cohen, J.C. Egrie, P.H. Lai, F.K. Lin, T. Strickland, E. Watson, N. Stebbing, Cold Spring Harb. Symp. Quant. Biol. 51 (1986) 693–702.
[11] J.C. Egrie, T.W. Strickland, J. Lane, K. Aoki, A.M. Cohen, R. Smalling, G. Trail, F.K. Lin, J.K. Browne, D.K. Hines, Immunobiology 172 (1986) 213–224.
[12] R.S. Rush, P.L. Derby, T.W. Strickland, M.F. Rohde, Anal. Chem. 65 (1993) 1834–1842.
[13] R.S. Rush, P.L. Derby, D.M. Smith, C. Merry, G. Rogers, M.F. Rohde, V. Katta, Anal. Chem. 67 (1995) 1442–1452.
[14] M.N. Fukuda, H. Sasaki, L. Lopez, M. Fukuda, Blood 73 (1989) 84–89.
[15] J.L. Spivak, B.B. Hogans, Blood 73 (1989) 90–99.
[16] L.O. Narhi, T. Arakawa, K.H. Aoki, R. Elmore, M.F. Rohde, T. Boone, T.W. Strickland, J. Biol. Chem. 266 (1991) 23022–23026.
[17] R. Westermeier, Electrophoresis in Practice, 4th ed., Wiley-VCH, Weinheim, 2005.
[18] P.G. Righetti, Immobilized pH Gradients: Theory and Methodology, vol. 20, Elsevier Biomedical Press, Amsterdam, 1990.

Scientific Publishing House
offers
free of charge publication

of current academic research papers, Bachelor´s Theses, Master's Theses, Dissertations or Scientific Monographs

If you have written a thesis which satisfies high content as well as formal demands, and you are interested in a remunerated publication of your work, please send an e-mail with some initial information about yourself and your work to *info@vdm-publishing-house.com.*

Our editorial office will get in touch with you shortly.

VDM Publishing House Ltd.
Meldrum Court 17.
Beau Bassin
Mauritius
www.vdm-publishing-house.com

Printed by Books on Demand GmbH, Norderstedt / Germany